YO-CFT-284

TURTLES
of **BORNEO**
AND PENINSULAR MALAYSIA

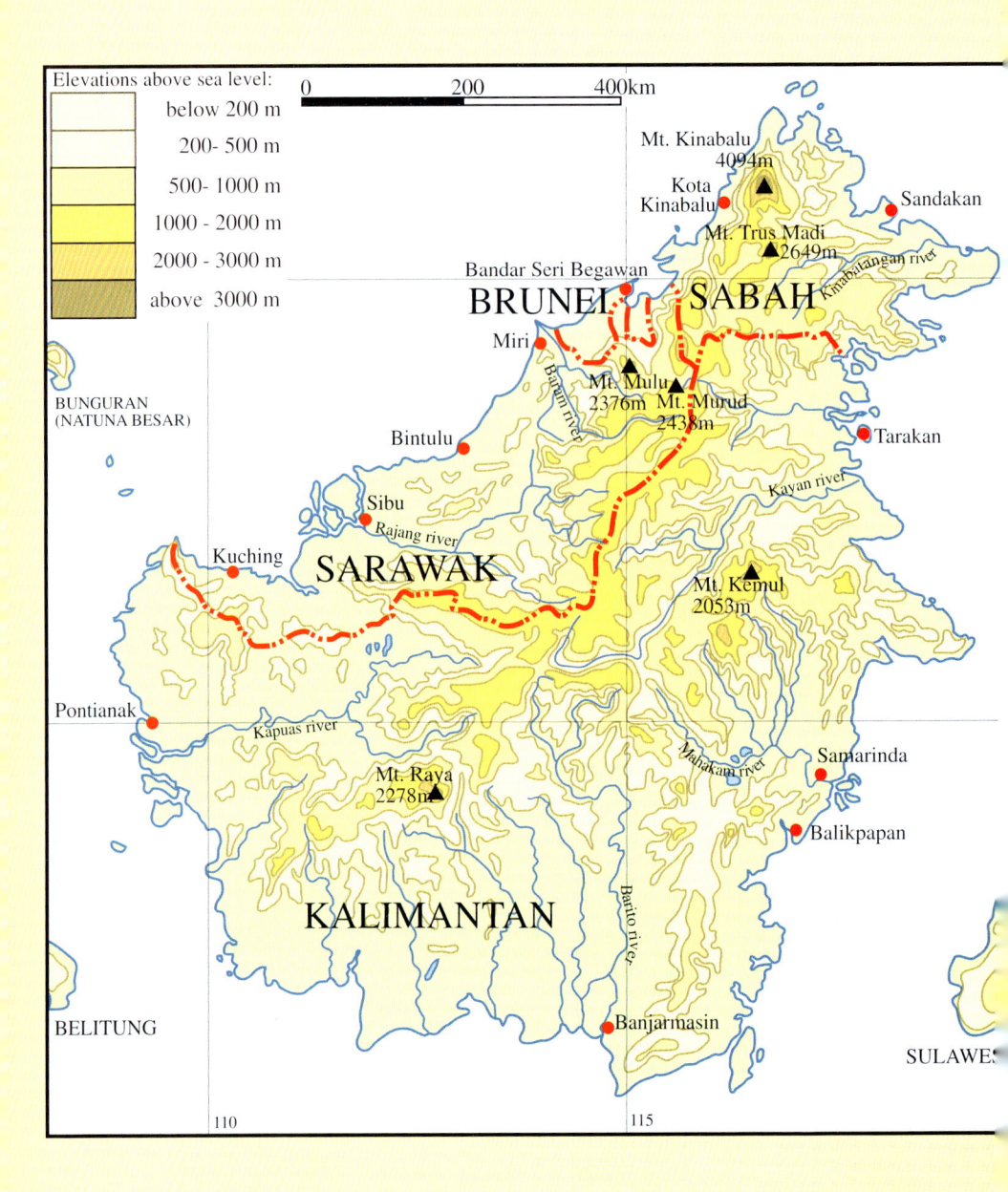

Map of Borneo.

TURTLES *of* BORNEO
AND PENINSULAR MALAYSIA

Lim Boo Liat and Indraneil Das

Foreword by
Y.B. Datuk Chong Kah Kiat, J.P.

Photographs by
C.L. Chan, Arthur Y.C. Chung, Christian Gönner,
Hans P. Hazebroek, Jason Isley, Ulrich Manthey,
Nicolas Pilcher, W.M. Poon, Cede Prudente,
Francis Seow-Choen, Dionysius Sharma,
Tan Heok Hui, Yong Hoi Sen, and the authors

Natural History Publications (Borneo)
Kota Kinabalu

1999

In memory of

Professor John Leonard Harrison
A Pioneer in Malaysian Zoology
1917–1972

Published by

NATURAL HISTORY PUBLICATIONS (BORNEO) SDN. BHD.
(Company No. 216807-X)
A913, 9th Floor, Wisma Merdeka,
P.O. Box 13908,
88846 Kota Kinabalu, Sabah, Malaysia.
Tel: 6088-233098 Fax: 6088-240768
e-mail: chewlun@tm.net.my

Copyright © 1999 Natural History Publications (Borneo) Sdn. Bhd.

All rights reserved. No part of this publication may be reproduced, stored in a retrieval system, or transmitted in any form or by any means, electronic, mechanical, photocopying, recording, or otherwise, without the prior permission of the copyright owner.

First published 1999.

Turtles of Borneo and Peninsular Malaysia
  Lim Boo Liat and Indraneil Das

Contents page: River Terrapin (*Batagur baska*). (Photo: Indraneil Das)
Facing Foreword: Shuk Mun with Asian Brown Tortoise (*Manouria emys*).
    (Photo: C.L. Chan)
Facing Preface: Asian Brown Tortoise (*Manouria emys*). (Photo: Indraneil Das)

Design and layout by C.L. Chan.

Perpustakaan Negara Malaysia Cataloguing-in-Publication Data

Lim, Boo Liat
    Turtles of Borneo and Peninsular Malaysia / Lim Boo
    Liat and Indraneil Das ; with photographs by C.L.
    Chan ... [et al.] ; line illustrations by Yong Ket Hun.
    Bibliography; p. 121
    ISBN 983-812-039-1
    1. Turtles—Borneo. 2. Turtles—Semenanjung Malaysia.
    I. Das, Indraneil. II. Title.
    597.92

Printed in Malaysia.

Contents

Foreword   vii
Preface   ix

1   *Introduction*   1
2   *Key to species*   11
3   *Dermochelyidae* Leatherback Sea Turtle   15
4   *Cheloniidae* Sea Turtles   21
5   *Trionychidae* Softshell Turtles   37
6   *Bataguridae* Asian Hardshell Turtles   53
7   *Emydidae* American Hardshell Turtles   93
8   *Testudinidae* Land Tortoises   99
9   *Conservation*   113

Acknowledgements   116
Glossary   118
References   121
Index   138

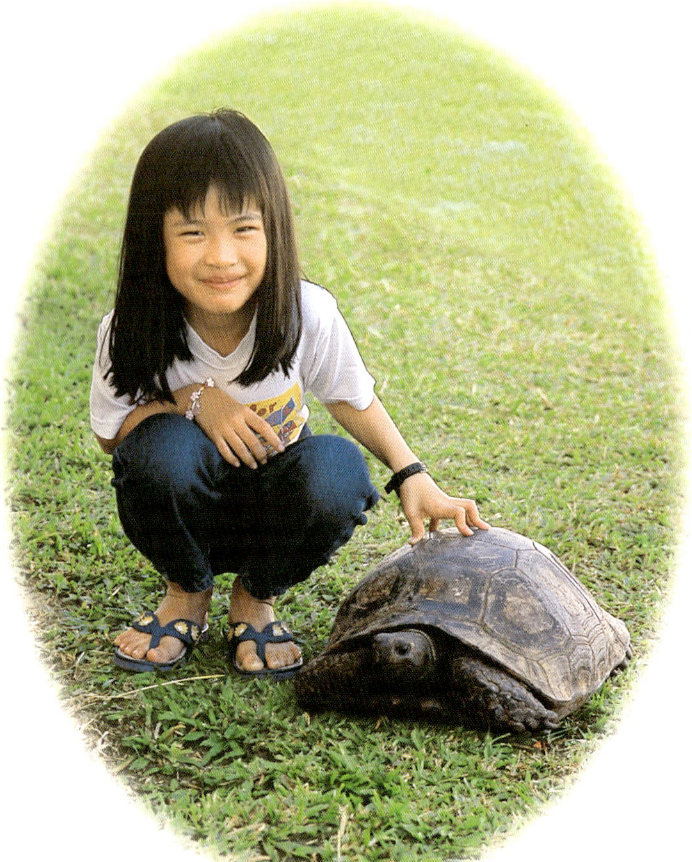

Foreword

by

Y.B. Datuk Chong Kah Kiat, J.P.
Minister of Tourism Development, Environment, Science and Technology, Sabah

Turtles have always managed to capture the imagination and interest of people and cultures. Sadly, though, not enough has been documented of these magnificent creatures in the Malay archipelago. While we hear of the wonders of Giant Leatherbacks congregating upon tropical shores and the plight of turtles dragged up together with fish catches, there is altogether too little available that presents to us the fascination of these animals as a group—few people have the opportunity to appreciate the overall diversity of a group of animals that has transcended the barriers between sea and land, represented by marvellous forms in Borneo and Peninsular Malaysia, swimming the oceans endlessly and roaming the great forests.

To say that this account helps the popular understanding and is another step forward in conservation is only a part of the achievement of the present book. It is written by two specialists who love turtles and whose combined efforts have brought out this overview of diversity, beauty and natural history of turtles.

From stories of the tortoise and the hare, to expressions such as turning turtle, and a universal amazement at how some turtles quite circumnavigate our planet, these animals have always held substantial human interest.

Now read on.

Chong Kah Kiat, J.P.

Indraneil Das

Preface

by

Joseph Pounis Guntavid
Director, Sabah Museum

The presentation of an account of turtles in the present format is a pleasant task. The Sabah Museum has been able to maintain its interest in documenting various aspects of Sabah's natural history and this has inevitably led to interaction with various specialists, including Dr Lim Boo Liat and Dr Indraneil Das. It is therefore a great delight that our holdings of specimens have been of use to them and one of the products of their research should be the present book.

Our interest in documentation is not restricted to the specialist research papers put out on specific aspects, but very much includes well-illustrated accounts such as here, which greatly help in popular understanding. Here must be mentioned specially the fortuitous presence of *both* biologists and a fine publisher who is truly interested in the natural history of our area.

Of turtles, this book reveals, in the simplest way, the many forms that not everyone gets to see or to read about in popular channels. One can only wish that—as is the case with turtles, so finely demonstrated—more of this region's natural history will be made available with the interest of specialists. Malaysia's natural heritage is a vast and priceless one, and making information more accessible is a vital step in the difficult process of management and conservation.

Joseph Pounis Guntavid

F. Seow-Choen

Chapter 1

INTRODUCTION

Turtles, tortoises and terrapins belong to the Order Chelonii (formerly Testudines) which include the terrestrial, freshwater, brackish water and marine forms. The Order is divided into two Suborders: Cryptodira and Pleurodira. Members of Crytodira have short necks, that are generally retractile inside the shells (exceptions include the marine turtles). Turtles belonging to Pleurodira have long, non-retractable necks that are instead withdrawn and folded to one side. In the region covered by this work (Borneo and Peninsular Malaysia), all turtles are cryptodirans. There are 12 living families and over 260 living species in the world, of which 25 species (including two exotics) have been recorded as occurring on Borneo and Peninsular Malaysia.

Turtles, along with lizards, snakes, crocodilians and a bizarre reptile from some of the islands off New Zealand, called the tuatara, constitute the group reptiles. There is evidence that these groups are not particularly closely related, some, such as the crocodiles and alligators being closer to birds than they are to other reptiles. Turtles are surely the most distinctive among the living reptiles, with a shell that incorporates the ribs, vertebrae, pectoral girdle and skin, giving protection to the body. No living turtle species possesses teeth (some of the extinct forms that we know from fossils did) and their skulls are without opening on the sides. The shell is composed of two major

---

(Opposite). The Aldabra Tortoise (*Dipsochelys gigantea*) from the Seychelles is one of the largest turtles in the world.

parts, the carapace (or upper shell), and the plastron (or lower shell), the two joined laterally by a narrow part laterally, called the bridge. Most turtles (except softshell turtles and the Leatherback Sea Turtle, *Dermochelys coriacea*) have horny (the same material as nails and horns) scales covering the shell, giving them additional protection. These scales are usually of great importance in identification of individual species, and, on the carapace, include a single row of a central series (the vertebrals) that follow the single nuchal scale behind the neck region, an intermediate series (the costals), and a peripheral series (the marginals). On the plastron, shell scales are paired, and include the gular, humeral, pectoral, abdominal, femoral and anal. The shape and texture of the carapace varies according to the ecological habits of the species, the land-dwelling forms showing a relatively more domed, often rugose shell, while aquatic species have a more flattened, smooth and streamlined shell. The carapace, and particularly the plastron, may not always be rigidly fixed, and in some species, are capable of movement (as might be needed for delivering large eggs, or wedging in narrow crevices).

The earliest turtles appeared during the Late Triassic, 230 million years before present, and the origin of the group remains largely unresolved. These early turtles are imperfectly known from partially preserved shells, skulls and other skeletal remains, from fossiliferous rocks of Germany. They had teeth on the bones of the palate, and distinct shells, little different in basic structure from the types to be seen in living turtles. The two major groups in which living turtles are divided are the Pleurodirans (the side-necked turtles, which bend their necks sideways under the front of the carapace) and the Cryptodirans (the vertical-necked turtles, which bend their necks into a vertical sigmoid curve; this group, however includes sea turtles that have a reduced ability to withdraw their head). All turtles from Borneo and Peninsular Malaysia belong to the latter group, which is thought to have evolved more recently. Pleurodirans have a distribution in the southern continents (South America, Africa south of the Sahara and Australasia), while Cryptodirans are more typically found in northern latitudes, although many species have invaded south Africa and South America, coexisting with the Pleurodirans. Fossil records reveal that Pleurodirans were once very widespread,

Fossil records reveal that Pleurodirans were once very widespread, and have probably been outcompeted by the more recently evolved Cryptodirans.

Turtles are among the most long-lived of the vertebrates, some of the largest species (such as those from the oceanic islands of Galapagos and Aldabra) attaining over a century, especially in captivity. Wild turtles, however, are likely to have shorter life spans, and freshwater turtles probably live for only a few decades. Turtles have an extensive behavioural repertoire, virtually unmatched by other reptiles. Many are social, basking or feeding in groups. Some (such as the Ridley Sea Turtles of the genus *Lepidochelys*) form dense nesting aggregations that cover virtually every square meter of sea beach during the nesting season. Several turtles are known to produce sounds, especially during courtship, including grunts, gasps, bellows and even chicken-like clucks. When threatened, some species are known to utter a deep hiss, others growl or croak. Sea turtles are highly aquatic, some species, such as the Leatherback making impressive migrations from their tropical nesting beaches to the polar

Close-up of scute from the shell of a Black Marsh Turtle (*Siebenrockiella crassicollis*), showing the growth rings.

characterised by flippers, which are forelimbs modified to look like the wings of birds, through the fusion of digits. Tortoises are poor swimmers, sometimes colonising islands by floating for short distances between islands. Their gait on land is slow, but the powerful, club-shaped hind legs and columnar forelegs can raise their bodies well off the ground. Freshwater turtles of some species are fast swimmers; others are known to bottom-walk, being inhabitants of stagnant or slow-moving waterways. The more aquatic ones show extensive webbing on their feet, while the terrestrial turtles lack webs.

All turtles whose nesting habits are known produce eggs, and with one exception (the Northern Long-necked Turtle, *Chelodina rugosa*, from Australia, whose eggs are laid underwater), are laid on land. Some of the land tortoises produce a single egg; others, such as the sea turtles, can lay over 200 at a time, and several times during a single nesting season. Eggs are generally laid in holes excavated underground with the help of the hindlimbs, the Asian Brown Tortoise (*Manouria emys*) differing in making a leaf-mound nest in which its large clutch of eggs is laid. For most species, sex is determined by the temperature of incubation of eggs. Softshell turtles (Family: Trionychidae), on the other hand, probably have the genotypic sex determination mechanism. Incubation period (or the time taken for the young turtles to emerge from the egg), is, to some extant dependent on the temperature at which the eggs develop, and may range from a few weeks to as long as a year and a half, as in some giant tortoises. In general, emergence of baby turtles is synchronized with the season of plenty (such as the wet months), when food in

(Opposite). A clear forest stream, home for many freshwater turtles, in Peninsular Malaysia.

the form of plant and animal life (such as insects) is abundant. There is no parental care once the eggs are deposited (the exception being the Asian Brown Tortoise, the females of which may stay a few days after completing nesting, pushing away potential intruders from their nests). Hatchlings of sea turtles and most other water turtles scramble towards water immediately upon emergence, in the case of sea turtles, returning to dry land only years later. Even then, it is only the females which will ascend the beach, while males linger in the shallow coastal waters.

Although turtles don't seem to have the public relations problem with people that crocodiles, snakes and large lizards seem to suffer from (presumably for lacking fangs or venom, they are as a whole considered "safe"), many species of turtles are in danger of extinction caused by humans. Almost throughout the world, turtles are eaten as either a proteinaceous food or for the purported therapeutic value of their body parts, such as meat, gall bladder, fat, unshelled eggs, cartilage, and bones that are used for the manufacture of drugs. The incredibly large number of freshwater turtles and tortoises from all over south-east Asia that land up in the markets of China is a particular cause for concern, as several species found in these markets are unknown to science, and therefore, nothing is known of their biology and conservation status. Large numbers of turtles are also collected from the wild for supply to a growing pet trade industry in the West. Even in many cultures that do not eat turtle meat, the harvesting of turtle eggs may be an acceptable activity, and on some sea turtle nesting beaches, virtually every egg laid are collected for consumption or sale. Other threats to the survival of wild turtles include the modification of their natural habitats. For instance, dams built on a river has the potential to change a running water ecosystem into a standing one, which, over time, will affect the ecology of the waterway, causing the extinction of species that are adapted to living in flowing waters. Pesticide and herbicide spill-offs from agricultural fields into ponds and rivers is another human induced threat to turtles and other aquatic organisms. Some pesticides are known to cause thinning of egg-shells in birds, leading to breakage during development. More specific conservation issues have been discussed in Chapter 9.

Turtles play a number of useful functions in the ecosystem. Many are feeders of carrion, often scavenging far from their typical haunts, and thereby help in the release of locked-up nutrients. Other turtles feed on organisms that cause diseases in humans, such as schistasomiasis, which is spread by a snail, and malaria and filaria, that are transmitted by mosquitoes; several aquatic turtles are specialised snail eaters (complete with enlarged jaw muscles for breaking the hard shells of snails) or feeders of mosquito larvae. By eating water weeds, such as the introduced water hyacinth (*Eichhornia crassipes*), water turtles help maintain open waters, and thereby in the control of insects (such as mosquitoes) that breed in stagnant waters. In turn, turtles are food for other animals, from large

Introduction

A juvenile Asian Softshell Turtle (*Amyda cartilaginea*), showing a skin-clad shell and nostrils set on a proboscis.

fishes to waterbirds, monitor lizards, otters and crocodiles, not to mention many aboriginal people throughout the world.

Finally, turtles are important to many local cultures, and several tribes consider the turtle as a totem, eating of the totemic animal being taboo (forbidden) for the tribe members. For others, bringing a turtle to a temple is considered a good deed, that brings not only good luck, but also a better after life.

This work described all species of native and introduced turtles and tortoises recorded from Borneo (including the Malaysian states of Sabah, Sarawak, as well as Brunei Darussalam and Indonesia's Kalimantan Province), besides Peninsular (West) Malaysia. A few of these spill over to Singapore, for which reason we have covered this island state as well. The primary aim of this work is to enable the lay public to identify species encountered in the field, and more generally, to create an interest in these animals. We use the word 'turtle' for the whole group, although in general English usage, turtle is the marine animal, tortoise the terrestrial animal, and terrapin, the freshwater one. Contrary to traditional usage, several terrapins have relatives that spend their entire life on dry land and are unable to swim, and a few tortoises prefer wet areas. The River Terrapin (*Batagur baska*) typically inhabits brackish, even coastal, waters, and generally avoids freshwaters.

We have tried to use external characteristics (such as size, colour and morphological characters), whenever possible as identification characters to define species, although for familial or generic identification, internal characters need to be studied. Each species account contains the English, the current scientific and vernacular names, the primary citation, besides data on morphological characteristics (including size, colour and sexual dimorphism), distribution, biology (including habitats preferred, diet, behaviour and reproduction) and conservation status. A glossary of technical terms used can be found at the back of the book.

CHECKLIST OF TURTLES OF BORNEO AND PENINSULAR MALAYSIA

(species marked with an asterisk are exotics that are established in the region)

DERMOCHELYIDAE
1. Leatherback Sea Turtle *Dermochelys coriacea* (Vandelli, 1761)

CHELONIIDAE
2. Loggerhead Sea Turtle *Caretta caretta* (Linnaeus, 1758)
3. Green Turtle *Chelonia mydas* (Linnaeus, 1758)
4. Hawksbill Sea Turtle *Eretmochelys imbricata* (Linnaeus, 1766)
5. Olive Ridley Sea Turtle *Lepidochelys olivacea* (Eschscholtz, 1829)

TRIONYCHIDAE
6. Asian Softshell Turtle *Amyda cartilaginea* (Boddaert, 1770)
7. Narrow-headed Softshell Turtle *Chitra indica* (Gray in: Griffith & Pidgeon, 1831) species-complex

8. Malayan Softshell Turtle *Dogania subplana* (Geoffroy-Saint Hillaire, 1809)
9. Asian Giant Softshell Turtle *Pelochelys cantorii* Gray, 1864
10. Chinese Softshell Turtle *Pelodiscus sinensis* (Wiegmann, 1835)*

BATAGURIDAE
11. River Terrapin *Batagur baska* (Gray in: Gray & Hardwicke, 1830)
12. Painted Terrapin *Callagur borneoensis* (Schlegel & Müller, 1844)
13. Malayan Box Turtle *Cuora amboinensis* (Daudin, 1801)
14. Asian Leaf Turtle *Cyclemys dentata* (Gray, 1831)
15. Orange-headed Temple Turtle *Heosemys grandis* (Gray, 1860)
16. Spiny Hill Turtle *Heosemys spinosa* (Gray, 1831)
17. Yellow-headed Temple Turtle *Hieremys annandalei* (Boulenger, 1903)
18. Malayan Snail-eating Turtle *Malayemys subtrijuga* (Schlegel & Müller, 1844)
19. Malayan Flat-shelled Turtle *Notochelys platynota* (Gray, 1834)
20. Asian Giant Turtle *Orlitia borneensis* Gray, 1873
21. Black Pond Turtle *Siebenrockiella crassicollis* (Gray, 1831)

EMYDIDAE
22. Red-eared Slider *Trachemys scripta* (Schoepff, 1792)*

TESTUDINIDAE
23. Elongated Tortoise *Indotestudo elongata* (Blyth, 1853)
24. Asian Brown Tortoise *Manouria emys* (Schlegel & Müller in: Temminck, 1844)
25. Impressed Tortoise *Manouria impressa* (Günther, 1882)

Note: The record of *Geoemyda spengleri* from Borneo is in error.

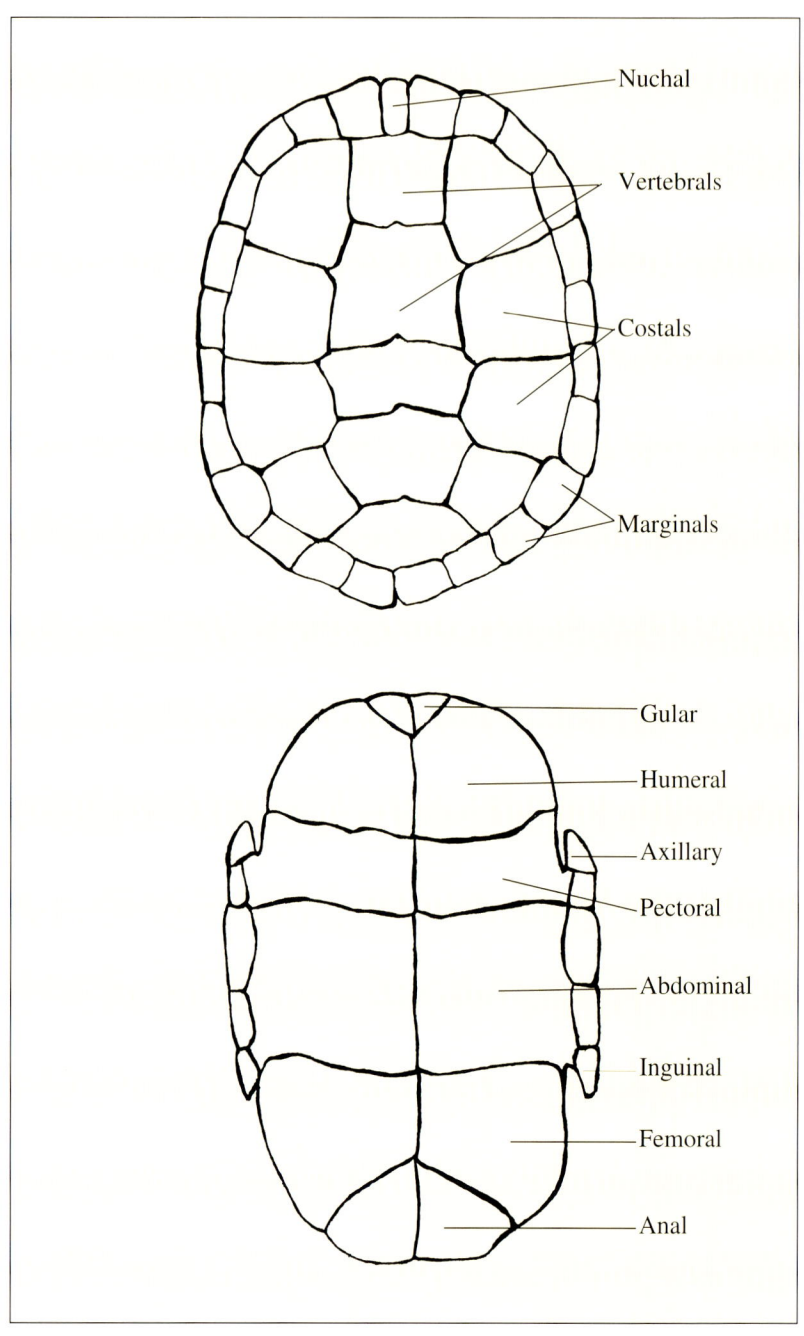

Dorsal (above) and ventral (below) views of the carapace and plastron of a typical turtle. (Line illustrations by Indraneil Das).

Chapter 2

KEY TO SPECIES

1. Forelimbs flipper-like; head non-retractable; marine 2
   Forelimbs not flipper-like head retractable; estuarine, freshwater or terrestrial 6

2. Shell covered with leathery shell (small scales present in juveniles only), with distinct ridges on carapace and plastron, shell tapering posteriorly to a narrow point .... *Dermochelys coriacea* (p. 15)
   Shell covered with scutes, lacking ridges, posteriorly more or less rounded 3

3. Four pairs of costals on carapace 4
   Five or more pairs of costals on carapace 5

4. Carapace scutes juxtaposed; snout blunt; two frontoparietals on forehead; size to 1400 mm .... *Chelonia mydas* (p. 24)
   Carapace scutes imbricating; snout tapering to form a beak; four frontoparietals on forehead; size to 1140 mm .... *Eretmochelys imbricata* (p. 30)

5. Colour reddish-brown; bridge with three inframarginals; size to 1200 mm .... *Caretta caretta* (p. 22)
   Colour greyish-olive; bridge with four inframarginals; size to 750 mm .... *Lepidochelys olivacea* (p. 34)

6. Hindlimbs club-shaped, with free digits 7
   Hindlimbs not club-shaped; digits sometimes free 9

11

7. Supracaudas fused; carapace pale with dark blotches; size to 360 mm .... *Indotestudo elongata* (p. 99)
   Subcaudals divided; carapace dark brown or yellowish-brown, lacking darker blotches 8

8. Several large pointed tubercles on each thigh; carapace dark brown; posterior marginals weakly serrated; size to 560 mm .... *Manouria emys* (p. 103)
   A single large tubercle on each thigh; carapace yellowish-brown to brown; posterior marginals distinctly serrated; size to 302 mm .... *Manouria impressa* (p. 110)

9. Shell covered with skin; snout tube-like 10
   Shell covered with scutes; snout not tube-like 14

10. Snout shorter than eye diameter; olive-grey shell, sometimes with irregular specklings; size to 1500 mm .... *Pelochelys cantorii* (p. 46)
    Snout as long as eye diameter; shell variously coloured 11

11. Triturating surface of jaws narrow and sharp; neck and carapace with a linear pattern; size to 1220 mm .... *Chitra indica* species-complex (p. 41)
    Triturating surface of jaws broad and flat; neck and carapace without linear pattern 12

12. Four plastral callosities; carapace more or less with straight sides; proboscis strongly-downturned; carapace with a dark median stripe; cheeks reddish-brown; size to 310 mm .... *Dogania subplana* (p. 43)
    Over four plastral callosities; carapace rounded or oval; proboscis straight; carapace without a dark median stripe; cheeks not reddish-brown 13

13. Seven plastral callosities; head and neck with fine black lines; size to 165 mm .... *Pelodiscus sinensis* (introduced) (p. 49)
    Five plastral callosities; yellow or cream spots on head; size to 830 mm .... *Amyda cartilaginea* (p. 38)

14. Forearms four clawed; carapace of adults smooth, with seams hardly discernable; size to 560 mm .... *Batagur baska* (p. 54)
    Forearms five clawed; carapace smooth or keeled 15

15. Plastron with a free moving hinge 16
    Plastron rigid 17

16. Head with pale yellow or orange stripes; plastral scute with dark blotches; size to 250 mm .... *Cuora amboinensis* (p. 63)
    Head various, but never with yellow or orange stripes; plastral scutes with radiating black lines or unpatterned; size to 240 mm .... *Cyclemys dentata* (p. 66)

17. Carapace with black longitudinal stripes; forehead of breeding males with an orange or red patch; size to 600 mm .... *Callagur borneoensis* (p. 58)
    Carapace without black longitudinal stripes; forehead variously coloured 18

18. Carapace with six or seven vertebrals; size to 360 mm .... *Notochelys platynota* (p. 81)
    Carapace with five vertebrals 19

19. Carapace unpatterned black 20
    Carapace variously coloured, but never black 21

20. Juvenile shells tricarinate (keels faint in adults); forelimbs with enlarged scales; size to 200 mm .... *Siebenrockiella crassicollis* (p. 90)
    Juvenile shells unicarinate (smooth in adults); forelimbs lack enlarged scales; size to 800 mm .... *Orlitia borneensis* (p. 85)

21. Head with bright yellow stripes; carapace tricarinate; size to 200 mm .... *Malayemys subtrijuga* (p. 79)
    Head without yellow stripes; carapace unicarinate 22

22. Posterior marginals strikingly serrated, especially in juveniles; digits weakly webbed; head greyish-brown with a yellow spot near tympanum; size to 225 mm .... *Heosemys spinosa* (p. 72)
    Posterior marginals weakly serrated; digits with extensive webbing; head brightly coloured 23

23. Sides of head with a bright red patch; size to 280 mm .... *Trachemys scripta elegans* (introduced) (p. 93)
    Sides of head lacking a red patch 24

24. Vertebral keel well defined; sides of head with orange or pink-mottled spots, best marked in juveniles; size to 480 mm .... *Heosemys grandis* (p. 70)
    Vertebral keel low; sides of head with yellow or cream stripes, best marked in juveniles; size to 506 mm .... *Hieremys annandalei* (p. 77)

Chapter 3

DERMOCHELYIDAE
LEATHERBACK SEA TURTLE

Remarkable in many aspects of its morphology and ecology, the family includes a single living species, the Leatherback Sea Turtle (*Dermochelys coriacea*), also the largest living turtle. Similar to sea turtles of the family Cheloniidae, this too is entirely marine, only the adult females coming ashore to nest. Instead of a scute-covered shell, adults have a dark, leathery skin covering a lyre-shaped body that tapers to a narrow point over the tail, and lack claws. The hatchling shell is covered with tiny scales that are lost during growth. Leatherbacks are specialist jellyfish-eaters. This is the only turtle that habitually ventures into temperate seas and dives deep in search of food.

Leatherback Sea Turtle

Dermochelys coriacea (Vandelli, 1761)
Malay *Penyu timbo, Penyu belimbing, Kamban, Kembau, Agal*

Testudo coriacea Vandelli, 1761. Rendus Hebd, Séanc. Acad. Sci., Paris 101(6): 2.
Derivation: Latin for leathery, in allusion to its leathery shell.

IDENTIFICATION. The Leatherback Turtle is the largest sea turtle in the world. The largest known specimen, which was washed up on the coast in Wales was 1800 mm in carapace length and weighed 916 kg. Apart from their colossal size, diagnostic features that help identify a Leatherback include lack of scales on adult shells (hatchlings have small polygonal scales), which is covered with leathery skin, and carapace with seven and plastron with five longitudinal ridges. The carapace has the appearance and texture of hard black rubber, narrowing to an acute point at the rear. Its eyelids close vertically and the upper jaw bears two toothlike projections that are flanked by deep cusps. Adults are dark brown or black above, either spotted or blotched with pale yellow or white, plastron paler. Juveniles are blackish-blue above, margins of limbs yellowish-cream; ventrum white with black markings. Males have concave plastra with a more or less depressed shell; posterior of shell tapering, with the tail, which is over double that of females, longer than the hindlimbs.

DISTRIBUTION. Cosmopolitan, in tropical, subtropical and temperate waters.

BIOLOGY. Leatherbacks are open-ocean turtles, migrating the furthest, reaching nearly the poles and are capable of deep dives. The longest migration recorded is 5900 km, and they are known to attain speeds of up to 10 km per hour. One individual completed a dive 1200 m, the deepest dive by a vertebrate. The body temperature of this turtle can go up to 18°C higher than that of the surrounding waters, primarily due to an insulating fat

Paper Land Sdn. Bhd.

Dermochelyidae / Dermochelys coriacea

Leatherback Sea Turtle (*Dermochelys coriacea*) nesting at Rantau Abang beach in Trengganu, Peninsular Malaysia. Adults lack scutes on their shells—instead, the shells are covered with leathery skin. The dark shell and limbs bears pale spots.

Leatherback Sea Turtle (*Dermochelys coriacea*). Hatchling. Note the tiny scales on the carapace, which will disappear as the turtle grows.

layer, a black shell that presumably absorbs solar radiation and a counter-current circulatory system in the fore- and hindlimbs to counter heat loss. It is believed that this temperature controlling mechanism enables this species to migrate to temperate seas in search of food deep down. They thrive on jellyfish (Schyphomedusidae), the diet incorporating other marine invertebrates such as plankton, crustaceans, symbiotic fish, medusae and ctenophores, that are associated with jellyfish. These turtles also feed on pyrosomas during deep dives, colonies of which are probably located by their bioluminescence. Maturity is estimated to be attained in five years, although some populations are reported to attain maturity as early as two years. The smallest nesting female measured 1400 mm in carapace length and weighed 300 kg.

At Rantau Abang beach, these turtles nest only along a 19 km stretch of beach, between May and September, with a peak during June and July. The total number of nesting females is estimated to be 600 per year, with each female nesting about six times a season. Typically, clutch size is between 50–140, and females may nest more than once during a nesting season, at intervals of 9–14 days. The eggs are spherical, softshelled, white with green flecks, measuring 48–54 mm. Generally, a few yolkless eggs are deposited over the real eggs in a nest. Hatching takes about 50–60 days. As with other marine turtles, the incubation temperature affects the sex of the hatchlings: temperatures in the range 27–28.75° C produce males, while those in the range 29.75–32° C result in females.

STATUS. The flesh of the Leatherback is dark and coarse and is seldom eaten. However, their eggs are sought after for their purported aphrodisiac and other medicinal properties. Oil from this species is used for fixing leaks in wooden boats. Leatherbacks are known to eat plastic bags floating in the sea, probably mistaking them for jellyfish. Other forms of marine pollution may be contributing to the decline of this species in recent years.

In Peninsula Malaysia, the Rantau Abang beach in Trengganu is famous for nesting Leatherbacks. Their yearly visits by nesting females to the beach provide a livelihood for the local people through the harvesting of eggs. The Rantau Abang beach attract both international and local tourists yearly to marvel at this phenomenon. During the nesting season, as many as 300 tourists may visit the beaches every night. However, there has been a drastic decline in nesting turtles, as reflected by the fall in the number of eggs, from 853,000 in 1956 to 42,000 in 1987. The species is listed in the IUCN Red List of Threatened Animals as 'Endangered', and protected under Appendix I of CITES.

Pulau Selingan, Turtle Islands Park, eastern Sabah, Malaysian Borneo.

Chapter 4

CHELONIIDAE
SEA TURTLES

Sea turtles include a small group of turtles that live all their lives in the sea, with only the adult females coming ashore to nest. Like the Leatherbacks, they have forelimbs shaped like flippers, and possess claws. The shell is oval or heart-shaped, is covered with large scutes throughout their lives, while the plastron is reduced. The neck is rather short, and cannot be retracted within the shell. They have a great variety of diets, from hard-bodied molluscs, to sponges and sea weeds.

There are four genera and six species of sea turtles. Those that occur in the waters of the region include: the Loggerhead Sea Turtle (*Caretta caretta*), Green Turtle (*Chelonia mydas*), the Hawksbill Sea Turtle (*Eretmochelys imbricata*), and the Olive Ridley Sea Turtle (*Lepidochelys olivacea*).

Of these, three species, the Green, Hawksbill and Olive Ridley, are known to nest along the shores of Peninsular Malaysia, and the first two nest along the coastline of Borneo. A fourth species, the Loggerhead, has not been known to nest in the region, but is an occasional visitor to West Malaysian waters, as indicated by individuals that are sometimes caught in trawling nets.

Sea turtles are easily recognized: they have flippers that are paddle-shaped forelimbs, used for swimming in a manner comparable to the wing-beats of birds. Sea turtles spend nearly their entire life in the sea. Mating occurs in shallow coastal waters, close to nesting beaches and females come ashore to lay eggs. The males generally never come ashore. A nesting female ascends the beach usually at night during high tide. Here, a site well above the high water mark is selected apparently after much deliberation and the eggs laid. Using her fore- and hindlimbs, the female excavates a depression (the 'body pit'), for the whole body to settle in. She then digs a deeper egg chamber into the moist sand with her hindlimbs and lays the eggs. When laying is completed, she covers the eggs, compacts the nest and just before she heads towards the sea, sometimes tries to hide evidence of nesting by scattering sand and pulling up beach vegetation all around for some distance. The clutch size varies between species as does the incubation period.

Loggerhead Sea Turtle
Caretta caretta (Linnaeus, 1758)
Malay *Penyu karah, Penyu kepala besar, Penyu sisek tempurong*

Testudo Caretta Linnaeus, 1758. *Syst. Nat.*, Ed. 10, 1: 197.
Derivation: Spanish for scutes (of turtles).

IDENTIFICATION. The Loggerhead Sea Turtle is named for its strikingly large head. Carapace has five pairs of juxtaposed costal scutes, first pair small; head massive; a single pair of prefrontal scales; bridge usually with three inframarginals; upper jaw hooked; carapace reddish-brown. Carapace length up to 1200 mm and weighs up to 110 kg. Because they are similar to the Olive Ridley, the two have been frequently confused in the literature. Adult males are smaller than females, have shorter plastra, more depressed and wider carapaces, and possess relatively longer tails that project out of the carapace rim.

DISTRIBUTION. Tropical and subtropical seas, from continental shelves, bays, lagoons and estuaries. Within the region, nesting occurs in Australia, Southern China and Japan.

BIOLOGY. The Loggerhead is a casual visitor to Peninsular Malaysia and Singapore, and are occasionally caught in trawl nets. Adults occur mainly in continental coastal waters, though juveniles may be associated with floating weed masses in the open sea. Diet comprises a variety of bottom-living organisms, such as molluscs, crabs, sea urchins, sponges and fishes. The Loggerhead has not been known to nest here, but extralimitally lay 23–178 spherical, papery or with leathery texture, eggs that measure 34.7–55.2 mm. The incubation period is 49–80 days. Hatchlings measure 34–55 mm.

STATUS. Because of its rarity in the seas off West Malaysia and Singapore, this species has been excluded from many faunal lists for the region. Listed in the IUCN Red List of Threatened Animals as 'Endangered', and included in Appendix 1 of CITES.

Loggerhead Sea Turtle (*Caretta caretta*). The reddish-brown carapace and extremely large head (adaptive for crushing molluscs) identify this species.

Green Turtle

Chelonia mydas (Linnaeus, 1758)
Malay *Penyu agar, Penyu hijau, Penyu pulau, Penyu emegit*

Testudo mydas Linnaeus, 1758. *Syst. Nat.*, Ed. 10, 1: 197.
Derivation: Greek for wet.

IDENTIFICATION. The Green Turtle is the most common sea turtle on Borneo and West Malaysia. Carapace with four non-overlapping coastal scutes; snout short and blunt, without a hook; distance from nostril to eye shorter than eye diameter; a single pair of prefrontal scales, four postorbital scales behind eye. Adults attain a carapace length of over 1400 mm and weigh up to 150 kg. There is a single claw on each limb. Shell smooth, brownish-olive, with spots or radiating streaks of light brown or black. Females usually more richly

Green Turtle (*Chelonia mydas*) resting on coral reefs, Layang Layang. The single pair of prefrontal scales and usually larger size help differentiate this species from the somewhat smaller Hawksbill (*Eretmochelys imbricata*), which has two pairs.

Cheloniidae / Chelonia mydas

C.L. Chan

C.L. Chan

(Above). Green Turtle (*Chelonia mydas*) nesting on Pulau Selingan, Turtle Islands Park, Sabah. (Below). A female Green Turtle laying eggs.

Turtles of Borneo and Peninsular Malaysia

Cheloniidae / Chelonia mydas

Green Turtle (*Chelonia mydas*) (Inset). Close-up of head.

27

pigmented. Juvenile carapace olive or dark brown. Limbs yellow-margined; plastron yellow. Adult males are smaller than females and possess a more tapered carapace, a large claw on each forelimb, and a relatively longer tail that project out of the carapace rim.

DISTRIBUTION. Cosmopolitan, occurring in tropical waters, especially around oceanic islands and coasts with wide sandy beaches. Other regional nesting beaches of importance include those in Indonesia, Myanmar, the Philippines and Thailand.

BIOLOGY. Adult Green Turtle are associated with shallow, calm coastal waters, 3–10 m deep, especially sandy areas with seagrass beds. Besides sea grasses, they also eat algae, although small animals may be incidentally ingested. Juveniles are carnivorous, feeding on aquatic invertebrates as they raft on sargassum. Adult Green Turtles have been observed resting on rocky crevices and in underwater caves.

Green Turtle (*Chelonia mydas*) hatchlings scrambling from their nest towards the sea. Predation is high during this stage of their life, enemies including birds, dogs and even fishes and crabs.

Cheloniidae / Chelonia mydas

Eggs of Green Turtle (*Chelonia mydas*). The incubation period of the eggs determine the sex of most turtles, and therefore, conservation programmes for these animals need to be planned carefully.

In Peninsular Malaysia, these turtles are found in the seas off the east coast, but they are more abundant off the East Malaysian States of Sabah and Sarawak and Brunei Darussalam, on Borneo. Green turtles mature in around 4–8 years, depending on the locality, and sexual maturity is earlier in captivity than in the wild. Breeding cycles are well marked, females returning to the same breeding beach after an interval of 2–3 years. Nests are usually on smooth-sloping sandy beaches. Eggs are spherical, with each clutch containing 98–172 eggs of diameter 41.4–42.1 mm that hatch about two months later. A female may nest 6–7, and sometimes, up to 11 times in a nesting season. The hatchlings are bluish-black in colour, and measure 44–59 mm.

STATUS. The English name is for the colour of the fat, historically in great demand for making turtle soup. Green Turtles have been heavily exploited for food for centuries by seafarers, the yearly harvest of eggs, particularly from some of the islands off East Malaysia, being estimated at over a million. The meat of the turtle is eaten as well. The species is listed in the IUCN Red List of Threatened Animals as 'Endangered', and in Appendix 1 of CITES.

Hawksbill Sea Turtle

Eretmochelys imbricata (Linnaeus, 1766)
Malay *Penyu karah, Penyu lilin, Penyu sisek*

Testudo imbricata Linnaeus, 1766. *Syst. Nat.*, 12th Ed. 1: 350. Derivation: Latin for imbricate, for the overlapping scales on all but very old turtles.

IDENTIFICATION. The Hawksbill Turtle is distinguishable from all other sea turtles by its strong, narrow, hooked, bird-like beak; carapace with four pairs of costal scutes; and two pairs of prefrontals. This is a medium-sized turtle, with a carapace length up to 1140 mm. Limbs clawed. Scutes smooth, imbricating (juxtaposed in old individuals). Hatchlings dark grey below with a distinct keel in each bridge region. Adults marbled yellow and dark brown above, yellow below. Juveniles brown dorsally, and blackish-grey ventrally, head shields and limbs brown with yellow borders. Adult males are smaller than females, have concave plastra, relatively longer claws on the forelimbs, as well as longer tails than females that project out of the carapace rim.

(Opposite). Hawksbill Sea Turtle (*Eretmochelys imbricata*), Sipadan Island, Sabah. (Above). Close-up of head showing the characteristic bird-like beak and two pairs of prefrontals.

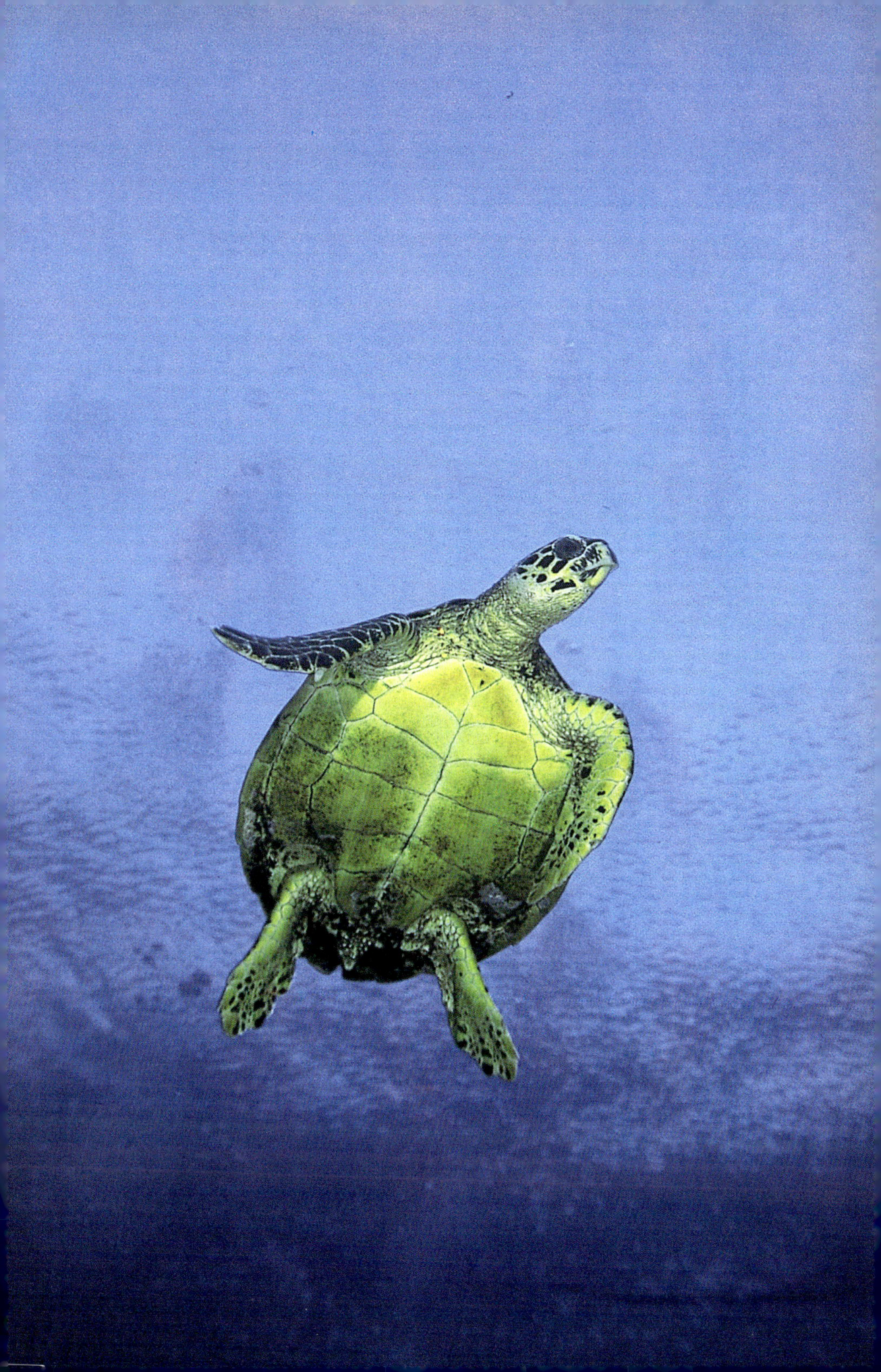

DISTRIBUTION. Cosmopolitan, in coral reefs, bays, estuaries and lagoons, in tropical and subtropical seas.

BIOLOGY. The Hawksbill lives in tropical and subtropical seas, generally in the vicinity of coral reefs. Nesting females can be found on both the east and west coasts of West Malaysia but are not as abundant as the Leatherback or Green Turtles. They are sometimes caught in trawling nets. The Hawksbill is omnivorous, specialising on sponges, and also taking molluscs, tunicates, sea anemones, crustaceans and algae. The hooked beak is probably adapted for prying out invertebrates from coral cervices. As this species is rather

Hawksbill Sea Turtle (*Eretmochelys imbricata*). A sponge-specialist, it can handle the sharp, glass-like structure that are present in these marine invertebrates. Hawksbills are threatened by demand for their scutes, needed for the manufacture of combs, curio boxes, jewellery and other tortoiseshell products.

Cheloniidae / Eretmochelys imbricata

An old Hawksbill Sea Turtle (*Eretmochelys imbricata*) nesting on Pulau Selingan, Turtle Islands Park, eastern Sabah.

rare in the region, little is known about the habits of the local population. The information provided here is primarily from extralimital nesting populations. Maturity is attained between 3–4.5 years. Nesting takes place on small beaches with coarse sand or gravel. A female may lay 2–4 nests a season, each comprising over 100 eggs. Nesting occurs over a period of 14–17 days, after which a female may nest again in about three years. Eggs are spherical, measuring 33–40 mm in diameter and incubation period is 52–57 days. Hatchlings measure 38–46 mm.

STATUS. Human exploitation of the Hawksbill is not confined to eggs and meat. The shell is highly sought after, its translucent scutes being used for making tortoiseshell items (such as combs, boxes and jewellery), and juveniles are stuffed for sale as curios. The flesh of the Hawksbill is not particularly palatable and the species is sometimes known to be poisonous. The Hawksbill is listed in the IUCN Red List of Threatened Animals as 'Critically Endangered' and the species is included in Appendix 1 of CITES.

Olive Ridley Sea Turtle
Lepidochelys olivacea (Eschscholtz, 1829)
Malay *Penyu lipas*

Chelonia olivacea Eschscholtz, 1829. *Zool. Atlas*, 1: 3, Pl. III.
Derivation: Latin for olive, in allusion to the colouration of the carapace.

IDENTIFICATION. The Olive Ridley is the world's smallest sea turtle, and is further distinguished by the possession of five to nine juxtaposed coastal scutes on the carapace. The almost circular carapace is slightly longer than broad. There are four inframarginals, each bearing pores, the functions of which are at present unknown. It attains a carapace length of 750 mm. Adults are uniformly olive or greenish-grey above, pale yellow below. Juveniles are dark brown or blackish-grey above, paler below. Hatchling carapace has three longitudinal keels while that of the adult is smooth. Adult males are about as large as the females, but possess relatively longer tails that project out of the carapace rim.

DISTRIBUTION. The Olive Ridley is widely distributed in the Indo-Pacific. Nesting takes place on the east coast of Peninsular Malaysia, but this species is not as common as the other sea turtles.

BIOLOGY. The Olive Ridley feeds on a variety of invertebrate prey, including crustaceans, molluscs, jellyfish, fish, as well as algae. Nesting beaches tend to be in the proximity of mangrove areas, and the Olive Ridley prefers broad, sandy beaches. The Olive Ridley differs from other marine turtles in two major features: body pits are not dug and nesting typically takes place in aggregations referred to as 'arribadas', with up to 150,000 females laying in a single night, as can be seen along the east coast of India. Although nesting is typically a nocturnal activity, females have been known to come

---

A nesting Olive Ridley Sea Turtle (*Lepidochelys olivacea*). This is the only sea turtle that nests in aggregations, referred to as 'arribadas'. Unlike the Loggerhead (*Caretta caretta*), the shell is olive or greenish-grey in colour and the head is not as enormous.

Cheloniidae / Lepidochelys olivacea

ashore to nest before sunset and after daybreak. The eggs are spherical, of diameter 36–39 mm and weigh 34–39 gm. Clutch size is 40–150 with an incubation period of 48–56 days. Hatchlings measure 35–45 mm. Nesting commences from December, lasting until April, and peaks between February and April.

STATUS. Egg harvesting from nesting beaches used to be an important economic activity in many areas in the past. In parts of its range, the flesh of the Olive Ridley is favoured, and large numbers of these turtles were once caught for food. At present, incidental take in trawling nets is a major threat to the species. Listed in the IUCN Red List of Threatened Animals as 'Endangered', and included in Appendix 1 of CITES.

Indraneil Das

A forest stream in the Batang Ai National Park in Sarawak, northern Borneo. Pristine habitats such as this are areas of greatest freshwater turtle diversity.

Chapter 5

TRIONYCHIDAE
SOFTSHELL TURTLES

The softshell turtles have been described as animated pancakes because of their smooth, flattened shells. These are highly specialised turtles that are almost never found far from water. They lack scutes on their shells, and are covered with a leathery skin with reduced shell bones. The snout is reminiscent of that of a pig, its limbs paddle-like, with each limb bearing just three claws. The head

An adult Malayan Softshell Turtle (*Dogania subplana*), characterised by its parallel-sided carapace with a dark median stripe. Softshell turtles are capable of underwater respiration, through absorption of dissolved oxygen.

and limbs can be retracted within the shell. The neck is relatively long, and these turtles tend to be aggressive, many capable of giving a painful bite. They are primarily carnivorous, although some plant matter may also be consumed. Scavenging is common.

Five genera, each with a single species, occur on Borneo and Peninsular Malaysia (and Singapore), including an introduced species. These are: the Malayan Softshell Turtle (*Amyda cartilaginea*), the Narrow-headed Softshell Turtle (an unknown member of the *Chitra indica* species complex), the Asian Softshell Turtle (*Dogania subplana*), the Asian Giant Softshell Turtle (*Pelochelys cantorii*), and the introduced Chinese Softshell Turtle (*Pelodiscus sinensis*).

Asian Softshell Turtle

Amyda cartilaginea (Boddaert, 1770)
Malay *Labi biasa, Bulus, Labi labi, Kuja*

Testudo cartilaginea Boddaert, 1770. *Brief Boddaert Kraakbeenige Schildpad*: 39.
Derivation: Latin for cartilage.

IDENTIFICATION. The commonest softshell turtle is the Asian Softshell Turtle, with a carapace length reaching 830 mm and weighing over 35 kg. The carapace is rounded in shape, with sides widest towards the posterior, not straight. They have distinct tubercles on the anterior carapace rim that extends in a single or double row. Juvenile carapace has numerous small tubercles and bony ridges, adults with smooth or some coarsely tuberculate (at rear) carapace. Proboscis is straight, not downturned. Dorsally brownish-black, grey or yellow below. Head and neck yellow spotted, 2–3 lines sometimes forming a V or arrow-shaped mark on forehead. Carapace occasionally with several poorly defined black eye-spots or dark markings. Males have relatively longer tails and outgrow females.

DISTRIBUTION. This species ranges from eastern and southern Myanmar, south to Thailand, Cambodia, Laos, Vietnam, West

Malaysia, Singapore, as well as the islands of Sumatra, Java, Borneo, Lombok and smaller associated islands.

BIOLOGY. This familiar softshell turtle occupies many different types of freshwater bodies, such as muddy rivers, ponds and irrigation canals, ascending hills up to low elevations. Like many members of the family, most of their time is spent buried in the mud or sand, with activities restricted to dawn or dusk. This is quite an aggressive species: large specimens kept in wire mesh cages are known to chew their way out with their powerful jaws and sharp beaks. Its diet comprises aquatic insects, crabs, prawns, fishes, carrion, as well as fruits and seeds. Females mature in 20 months, and nests are excavated close to waterbodies, between December and February. Three to four clutches may be produced per year, clutch size comprising 5–30 eggs. Eggs are spherical, brittle hardshelled, measuring 21–40 mm. Incubation period is between 61–140 days, with hatchlings emerging during the rains. These measure 37–49 mm in carapace length.

Asian Softshell Turtle (*Amyda cartilaginea*): close-up of head. The shell is more rounded than in the Malayan Softshell Turtle (*Dogania subplana*) and the head is pale spotted in all except very old individuals.

Status. This common softshell turtle is heavily exploited on account of the flavour of the flesh and for the belief by the ethnic Chinese that the flesh and the cartilage associated with the shell possess medicinal qualities. As a result, it fetches a price many times higher than pork, chicken or beef. There has apparently been a decline in the population of this species in Peninsular Malaysia in the last two decades, and softshell turtles are now imported from as far way as India by restauranteurs. In Singapore, this turtle has been reported as uncommon, and there have been attempts to farm the species for food. Apart from capture for food and medicine, habitat alternation, through developmental projects and water pollution are factors suspected to threaten this species. Listed by the IUCN Red List of Threatened Animals as 'Vulnerable'.

Narrow-headed Softshell Turtle

Chitra indica Gray in: Griffith & Pidgeon, 1831 species-complex
Malay *Labi bunga*

Trionyx indicus Gray in: Griffith & Pidgeon, 1831. *Class Reptilia* 9: 18.
Derivation: Pertaining to India, the type locality of *Chitra indica*.

Identification. The Narrow-headed Softshell Turtles of the genus *Chitra* contains some of the largest softshell turtle in the world, with carapace lengths reaching up to 1150 mm and weighing up to 120 kg. They are distinguished from other softshell turtles by their long narrow heads and colourful carapace patterns. The carapace rim smoothly joins the cartilaginous part to the skin of the neck; head tiny, with a miniscule proboscis; eyes situated close to the tip of the snout.

The Kanburi Narrow-headed Softshell Turtle, *Chitra chitra* has 2–4 scales on the lower arm; the Indian Narrow-headed Softshell Turtle, *C. indica*, has five or more scales. Dorsally brown with bright yellow, dark-edged stripes, the most striking of which is the inverted

Kanburi Narrow-headed Softshell Turtle (*Chitra chitra*), a Thai species, showing the brightly patterned, flat, circular carapace and a tiny head.

chevron-shaped mark on anterior of carapace and neck. Plastron white or pale pink. *Chitra chitra,* from Thailand attains a shell length of 1220 mm and probably even more, tipping the scale at 150 kg. Males outgrow females and have relatively longer tails.

DISTRIBUTION. The Narrow-headed Softshell Turtles of the genus *Chitra* are widespread in southern Asia (occurring in the rivers Indus, Ganga, Godavari, Padma, Mahanadi, and Coleroon, as well as in Pakistan, Nepal, India and Bangladesh). A related species, the Kanburi Narrow-headed Softshell Turtle (*Chitra chitra*), is restricted to the Mae Klong river basin of western Thailand at present; in the past it reportedly occurred in other river systems of central Thailand. Other localities in south-east Asia from where these giant turtles have been reported include Solo River in Java, Sumatra, Karimundjawa and the Ayeyarwady River of Myanmar. The identities of these turtles are at present unclear. The first West Malaysian specimen of these turtles *C. indica* was caught in the Kuala Tahan River, Pahang, in

1922. There are no recent records of these turtles from the Malay Peninsula.

BIOLOGY. There is no information on the feeding and breeding behaviour of the population from West Malaysia. The two known species of the genus (*Chitra chitra* and *C. indica*) inhabit sandy sections of medium or large rivers. Both are carnivorous, specializing on fishes and molluscs, and in captivity, the Indian species has been observed to ambush fish underwater, by burying itself in the sand with only the eyes exposed, and lunging at passing fishes. In northern India, breeding occurs in August. Nests are laid between end-August and mid-September. Eggs are brittle-shelled, spherical, measuring 25.4–28.2 mm, clutches containing 65–187 eggs. Incubation period is 40–70 days, hatchlings emerging between the middle and the end of October, and measure 29–33 mm. The Thai species nests between February to April, with nests excavated on sandbanks of rivers. Clutches comprise 60–117 spherical hardshelled eggs, measuring 34–40 mm. Hatchlings emerge after 62–66 days and measure 40–42 mm.

STATUS. The only specimen of this remarkable turtle from the Malay Peninsula was obtained from the Kuala Tahan River over 70 years ago. This suggests that the species is either extremely rare or occurs in inaccessible areas. It is absent from Singapore. This is one of the largest of the freshwater turtles in the region and is exploited for its flesh, large numbers are sold in the markets of India and Bangladesh, and till recently, it was exploited in Thailand as well. Developmental projects in and around rivers, including pollution and sand mining are possible threats to these turtles. *Chitra indica* is listed under Action Plan Rating 3 of the IUCN/SSC Tortoise and Freshwater Turtle Specialist Group. The IUCN Red List of Threatened Animals considers *C. chitra* as 'Critically Endangered'. These turtles are not listed by CITES. Their highly specialized habitat and dietary requirements are probably factors that threaten them.

REMARKS. The generic name is after the north Indian vernacular, meaning picture, for the striking carapace pattern. Since the Thai population has been shown to be specifically different from the ones in southern Asia, the identity of the West Malaysian population is of interest, and can be resolved only through the collection of new material. Although the type locality of *Chitra indica* is given as "India, fl. Ganges, Penang", it was subsequently restricted to Fatehgarh on the Ganga River in northern India. There have been no recent records from Penang (= Pulau Pinang, West Malaysia), although Farkas (1994) showed that a specimen at the Royal College of Surgeons of England, London (from Penang), that was destroyed during the Axis bombing of London during World War II, was part of the type series.

Malayan Softshell Turtle
Dogania subplana (Geoffroy-Saint Hillaire, 1809)
Malay *Labi melayu, Layi layi*

Trionyx subplanus Geoffroy-Saint Hillaire, 1809. *Ann. Mus. Hist. nat. Paris* 14: 11.
Derivation: Latin for dim beneath, for the dusky venter.

Malayan Softshell Turtle (*Dogania subplana*). Close-up of head, showing the pig-like snout and fleshy lips.

Indraneil Das

Trionychidae / Dogania subplana

IDENTIFICATION. The Malayan Softshell Turtle is easily recognized by its flattened (tubercles only on the posterior rim), straight-sided shell and a large head. The snout is relatively short, downturned, giving it a gnome-like appearance. It has two or three pairs of distinct ocelli and a black vertebral stripe. This is the smallest of the local softshells, with a maximum carapace length of just 310 mm. The adults are olive brown or yellowish-olive dorsally, cream or grey ventrally. The ocellus typically gets obscured during growth, while being more pronounced in juveniles. The juveniles are also more green in colour dorsally, and their plastra are usually white, with dorsum vaguely mottled or vermiculated. Adult males have slightly longer tails than adult females.

(Above). Malayan Softshell Turtle (*Dogania subplana*), juvenile.
(Opposite). An adult Malayan Softshell Turtle: note the straight-sided shell.

45

DISTRIBUTION. This species ranges from western Thailand (including the Mergui Archipelago), southwards through West Malaysia and Singapore, to Sumatra, Borneo and Java and their smaller, associated islands.

BIOLOGY. This species prefers slow flowing forest streams, and is more commonly found in the hills as well as foothills. On Langkawi Islands, they are known to aggregate in hill streams, although the function of this behaviour is not understood at present. They are excellent swimmers, and are omnivorous, feeding on fishes, prawns and other aquatic animals, besides algae and fruits that fall into the streams. The large head is suggestive of a diet of molluscs, as evidenced from other freshwater turtles with similarly enlarged heads. In captivity, it feeds on both fruits and vegetables. The turtle displays a crepuscular/nocturnal cycle, concealing itself between rocks on land. It has been suggested that the weak attachment of the carapace bones is an adaptation for hiding in such habitats. Clutches of 3–7 spherical eggs, of 22–31 mm diameter, are produced each year.

STATUS. The flesh of this softshell is held in high esteem, and large numbers are caught for food. This, coupled with habitat destruction and possibly competition with the introduced Chinese Softshell Turtle (*Pelodiscus sinensis*), are factors suspected to threaten the species. The species is not listed in the Appendices of CITES.

Asian Giant Softshell Turtle

Pelochelys cantorii Gray, 1864
Malay *Labi besar*

Pelochelys cantorii Gray, 1864. *Proc. Zool. Soc. London* 1864: 90. Derivation: For Theodore Cantor, the Danish naturalist who worked for the British East India Company.

IDENTIFICATION. One of the largest of freshwater turtles, this species exceeds 1500 mm in carapace length, tipping the scale at 200

Trionychidae / Pelochelys cantorii

kg. The shell is low and depressed, relatively more elongated in juveniles, oval in adult. Juveniles have numerous tubercles on the carapace and a low vertebral keel that disappears with age. The proboscis is extremely short and rounded. There are 3–4 small scales on the lower arm with small flaps of skin on the gular region. Dorsally, the carapace is olive or brown, with black speckles that are not very distinctive, this being a more sobre-coloured version of the Narrow-headed Softshell Turtles of the genus *Chitra*. The head is

Asian Giant Softshell Turtle (*Pelochelys cantorii*): adult. Diagnostic features include an unpatterned, depressed shell, and eyes situated close to the snout.

olive and the jaws white with darker markings; the outer surfaces of limbs and neck are olive or brown; the plastron yellow or cream. Juveniles are leaf-green with yellow spots dorsally.

DISTRIBUTION. The range of the species includes both coasts of India (although apparently not occuring in Sri Lanka), Bangladesh, Myanmar, Thailand, Malaysia, Indonesian Borneo (Kalimantan), southern China and Vietnam. Populations of *Pelochelys* from the north and south coasts of New Guinea are now assigned to two different species.

BIOLOGY. The species has been reported from the lower reaches of large rivers, coastal regions, as well as from ox-bow lakes and reservoirs. The Asian Giant Softshell Turtle is probably an ambush feeder specializing in fishes, although shrimps, crabs, molluscs, and sometimes, even aquatic plants, are consumed. Captives are known to accept vegetables and fruits. The turtle has been observed coming ashore to breed in the east coast of Peninsula Malaysia between February and March. The turtle is known to share the nesting beach with the Olive Ridley Sea Turtle in eastern India, the trionychid approaching the nesting beaches from both the river and the sea, though nesting is heavier on the riverside. Between 24–28 eggs are laid at a time on a river bank here, although Indian populations are known to nest on sandy sea beaches. The eggs are spherical with a diameter of about 30–35 mm, and have brittle hardshells.

STATUS. This species is incidentally caught while fishing for sharks and rays off Pulau Pinang. Similar to several other softshells, the meat of this species is favoured locally. The turtle is rare throughout its range and protected under Action Plan Rating 3 of the IUCN/SSC Tortoise and Freshwater Turtle Specialist Group. The biology of this large softshell turtle is poorly understood. It is not included in the Appendices of CITES. Hunting for food, and river development projects are suspected to be factors threatening the survival of these giant turtles.

REMARKS. Although specimens from West Malaysia in the past have been called *Pelochelys bibroni*, Webb (1995) restricted the type locality of *Pelochelys bibroni* to New Guinea, and showed that the name *Pelochelys cantorii* Gray, 1864 is available for the populations from the Asian mainland.

Chinese Softshell Turtle

Pelodiscus sinensis (Wiegmann, 1835)
Malay *Labi china*

Trionyx (Aspidonectes) sinensis Wiegmann, 1835. *Nova. Acta Acad. Caes. Leop. Carol.* 17: 189.
Derivation: From Latin, indicating China as the place of origin of the species.

IDENTIFICATION. A small softshell, carapace elliptical in adults, relatively more rounded in juveniles. The posterior part of the carapace has numerous tubercles, and the carapace margin is smooth

An adult Chinese Softshell Turtle (*Pelodiscus sinensis*). The back of the carapace is tuberculate, the front smooth, and the carapace is grey or olive, usually with many fine black marks.

Chinese Softshell Turtle (*Pelodiscus sinensis*): close-up of head.

except for a central tubercle above the neck. Carapace dorsum grey or olive, occasionally brown, with many fine black marks. The ventrum is orange with black markings in hatchings. With growth, the plastron becomes cream. Thin black lines radiate from the eye and the throat has several dark-bordered white spots. Straight carapace length to 165 mm, sometimes, even larger, and weight up to about 1 kg. Males have relatively longer tails that extend beyond the carapace margin, while females have a relatively higher shell.

Trionychidae / Pelodiscus sinensis

A hatchling Chinese Softshell Turtle (*Pelodiscus sinensis*).

DISTRIBUTION. The original distribution of the species includes Amur River of Russia, through Korea, China, Japan, Taiwan to northern Vietnam. It is now cultured in Malaysia, Vietnam, Indonesia and several other Asian countries, and is established in Hawaii.

BIOLOGY. The species occupies standing or slow-flowing bodies of water, such as ponds, reservoirs, marshes and rivers. Essentially nocturnal, they tend to bury themselves in the bottom of waterbodies. The diet typically consists of animal flesh, including fishes, crabs, snails, frogs, insects, in addition to carrion. Seeds of plants have also been found in their stomachs. Nests are excavated on mudbanks, and clutch size is 20–35, although a clutch of nine eggs is also known. The brittle hardshelled eggs have diameters of 16–20 mm. Eggs hatch after 50–60 days, and hatchlings measure 25–39 mm. Maturity is attained in 2–5 years in temperate regions, although in tropical areas, growth is more rapid, and maturity may be reached in about a year's time.

STATUS. Not listed in the IUCN Red List of Threatened Animals and not protected under CITES. Since the 1980s, this species has been farmed for food in Singapore and Thailand, in the latter country, mostly for export to markets in East Asia. Considered both delicious and of medicinal value, the demand for turtle meat and blood are high. In Singapore, there is a substantial introduced population while in Peninsular Malaysia, individuals are occasionally found in shallow ponds and canals in rural areas. While many farms may be self sufficient in terms of stock, the hazards of introduction of this exotic into relatively pristine areas (where they can compete with the local turtle species) remain.

Plastron of a Chinese Softshell Turtle (*Pelodiscus sinensis*) hatchling, showing the distinctive ventral pattern. With growth, the plastron becomes an unpatterned cream.

Chapter 6

BATAGURIDAE
ASIAN HARDSHELL TURTLES

As presently understood, the family is essentially Old World (a single genus occurs in central and South America) in distribution. Most of the diagnostic characters of the family are internal, requiring a study of the bones and cartilage of these turtles. In this section, we shall list only the superficial characters that separate these turtles from members of other families regionally: shell covered with scutes, hindlimbs not club-shaped, and may be free or webbed, nostrils not set on a proboscis and limbs with four or five claws. The head and limbs can be entirely retracted within the shell. These turtles inhabit forested areas or waterways, including ponds, streams, reservoirs, a few entering the sea. Some are good swimmers, others bottom-walkers, still others live their entire life on land. A few have semi-aquatic juvenile stages, becoming more terrestrial with adulthood. These turtles have a mixed diet, although some species are more carnivorous than others, and a few are specialists of one or two specific types of food.

Ten genera and 11 species occur in the region. These are: the River Terrapin (*Batagur baska*), the Painted Terrapin (*Callagur borneoensis*), the Malayan Box Turtle (*Cuora amboinensis*), the Asian Leaf Turtle (*Cyclemys dentata*), the Orange-headed Temple Turtle (*Heosemys grandis*), the Spiny Hill Turtle (*Heosemys spinosa*), the Yellow-headed Temple Turtle (*Hieremys annandalei*), the Malayan Snail-eating Turtle (*Malayemys subtrijuga*), the

Malayan Flat-shelled Turtle (*Notochelys platynota*), the Malayan Giant Turtle (*Orlitia borneensis*) and the Black Pond Turtle (*Siebenrockiella crassicollis*).

River Terrapin
Batagur baska (Gray in: Gray & Hardwicke, 1830)
Malay *Tuntong laut*

Emys baska Gray in: Gray & Hardwicke, 1830. "1830–1835". *Ill. Indian Zool.* 1: Pl. 75.
Derivation: Latin for bewitching, in reference to its beauty.

IDENTIFICATION. The River Terrapin is diagnosed by its large size (carapace length to 560 mm), with a domed and heavily buttressed carapace. It has a long plastron that is truncated anteriorly and notched posteriorly, triturating surfaces of jaws with two prominent ridges and four claws on each forelimb that have fully webbed digits. The carapace is smooth and moderately depressed. The head is rather

(Above). *Batagur baska*: hatchling. Note the serrated posterior carapace margin. (Opposite). The adults have an upturned snout, and unlike all other turtles, just four claws on its limbs.

small, the snout remarkable in being upturned and pointed. Juveniles have an interrupted ventral keel which disappears with growth. Carapace olive-grey, dark grey or brown, head similarly coloured, except being lighter on the sides; plastron unpatterned yellow. The head of males darken to jet black during the breeding season, when they also show a distinctive white ring around the iris during the breeding season.

DISTRIBUTION. The species ranges from eastern India and Bangladesh, through coastal areas of Myanmar, Thailand, Peninsular Malaysia, Sumatra (Indonesia), Cambodia and Vietnam. In West Malaysia, this species lives in large rivers, particularly in the Perak River and is also found in estuaries and mangrove areas. There are old records of the species from Singapore, where it is now extinct.

BIOLOGY. The River Terrapin occurs in the mouths of rivers under tidal influence and mangrove-dominated. Fruits of *Sonneratia*, a mangrove plant, are an important food item. Leaves, stems and other fruits are also consumed, besides molluscs, crustaceans and fishes.

In the Indian Sunderbans, a clutch comprising 19–37 eggs, measuring 68 × 40 mm was recorded. Incubation period ranges from

Turtles of Borneo and Peninsular Malaysia

Batagur baska adults. (Above). Male showing breeding colour, with intensely dark head. (Below). A non-breeding male. Its head is more brownish-olive.

Bataguridae / Batagur baska

Batagur baska adults. (Above). An intensely dark male, advertising its breeding condition. (Below). Resting underwater. An upturned snout enables it to breathe with most of its body underwater.

61–66 days. The breeding season in Peninsular Malaysia is from December to February, usually after the onset of monsoons. Nesting females migrate upriver to nest on sand banks. An average of 24 eggs are laid, with up to three nests laid in a year. Eggs are oblong, brittle hardshelled, measuring 66–40 mm and weighing about 64 gm. Hatchlings emerge in 66–68 days when incubated at 23–33° C.

STATUS. Collection of eggs of River Terrapins was an important economic activity for some coastal people of the region in the pre-World War II days. At present, threats to its survival include egg-collection, habitat destruction in the form of transformation of mangrove swamps into ponds for raising prawns, production of charcoal, harbours, in addition to aquatic pollution. Many terrapins are drowned in fishing nets or from injuries sustained from collision with boats. The species is listed as 'Endangered' in the IUCN Red Data Book and protected under Appendix I of CITES. It is listed under Action Plan Rating 1 of the IUCN/SSC Tortoise and Freshwater Turtle Specialist Group. The Department of Wildlife and National Parks has a recovery programme for the species at Bota Kanan (Perak), Bukit Pinang (Kedah), and Kuala Behrang (Trengganu). In each of these sites, eggs are collected and incubated and the hatchlings released.

Painted Terrapin

Callagur borneoensis (Schlegel & Müller, 1844)
Iban *Beluku*; Malay *Beluku, Biuku, Sutong, Tuntong sungei*

Emys borneoensis Schlegel & Müller, 1844. In: Temminck (ed.). *Verh. natuur. ges. Ned. Natuurk. Comm. Oost-Indie*: 30.
Derivation: Latin for inhabitant of Borneo.

IDENTIFICATION. The Painted Terrapin is perhaps the most colourful turtle in the region. Males in the breeding season have a

---

(Opposite above). An adult male of the Painted Terrapin (*Callagur borneoensis*) in its non-breeding colour. The three dark stripes on the carapace are diagnostic of the species. (Below). A male in partial breeding colour.

Lim Boo Liat

Dionysius Sharma

Turtles of Borneo and Peninsular Malaysia

Close-ups of the head of the Painted Terrapin (*Callagur borneoensis*). (Above). Male in breeding colour, with the black-edged red patch on the forehead. (Below). Male in non-breeding colour.

Bataguridae / Callagur borneoensis

Adult female (above) and hatchling (below) of the Painted Terrapin (*Callagur borneoensis*). This species is very similar to the River Terrapin (*Batagur baska*), but has five (not four) claws on each limb.

pale olive-grey carapace with three longitudinal black lines and black spots on the marginals. Head and upper part of neck bright white, with a black-edged red patch on forehead; jaws black. Lower neck, limbs and tail are grey. Outside the breeding season, colouration is more grey, with an orange forehead. Females are brown throughout the year, with only an indistinct orange forehead patch. In juveniles, the shell is flattened, and the vertebral keel distinct. The snout is pointed and upper jaw has a median ridge with denticulated edges of jaws. Forelimbs five clawed, digits broadly webbed. Carapace smooth with four pairs of costals. Adults reach a carapace length of 600 mm and weigh 25 kg. In addition, to the differences in colours, males have relatively longer and thicker tails that project out of the carapace rim.

DISTRIBUTION. The species inhabits the tidal section of rivers and estuaries of southern Thailand, Peninsular Malaysia, as well as Sumatra and Borneo.

BIOLOGY. The Painted Terrapin prefers large river systems and estuaries and is capable of tolerating up to 50% salinity. This species inhabits mangrove forests and feeds on mangrove fruits, leaves, as well as shellfish, such as clams. In captivity, it consumes vegetables and fruit. Mating takes place between January and February on the east coast of Peninsular Malaysia. Females migrate up to three kilometers to the nesting sites, that are on sand banks upriver or on sea beaches. In a season, 1–3 clutches of 15–25 large, elongated, flexible-shelled eggs are laid that measure 68–76 × 36–44 mm. These hatch after 85–98 days of incubation (at 31°C).

STATUS. The eggs of the Painted Terrapin are relished and its meat is in demand as well. On account of the belief that these beautiful turtles bring good luck to their owners, there is a market for these turtles as pets. Destruction of mangrove forests is a threat to the species. As a result of these pressures, populations of this species in West Malaysia are on the decline. The Wildlife and National Parks Department has a hatchery program in Kedah for the conservation of this species. This species is listed as 'Critically Endangered' in the IUCN Red List of Threatened Animals and included in Appendix II of CITES.

Malayan Box Turtle

Cuora amboinensis (Daudin, 1801)
Malay *Kura Katap, Kura kura, Kura kura patah*

Testudo amboinensis Daudin, 1801 "1802" *Hist. nat. Rept.* 2: 309.
Derivation: Latin, implying an inhabitant of Amboina, presently spelt Ambon, an island in Maluku, Indonesia.

IDENTIFICATION. This is the most common turtle on Borneo and in Peninsula Malaysia and Singapore. It is easily recognised by a well-developed plastral hinge, permitting the shell to be closed completely. The shell of the subspecies *kamaroma* from the region is high-domed and smooth, with a single vertebral keel in adult. Juveniles may show two additional lateral keels. Carapace margin smooth or slightly serrated; plastron rounded anteriorly and

Close-up of head of the Malayan Box Turtle (*Cuora amboinensis*). The bright yellowish-orange facial stripes are characteristic of this species.

Indraneil Das

Indraneil Das

posteriorly and attached to carapace by ligament. A transverse hinge at the pectoral-abdominal seam allows both lobes to be moved, permitting shell closure. Carapace is olive, brown or blackish-brown, plastron yellow or cream, with a single black blotch on each scute. Sides of head bear bright yellow or orange longitudinal stripes. Limbs olivaceous or yellow, digits entirely webbed. This is a medium sized turtle with a carapace length reaching 250 mm and weighing 1.5 kg. Adult males show a plastral concavity, while adult females have flat plastra.

DISTRIBUTION. The distribution of the species includes north-eastern India, Bangladesh, Myanmar, Thailand, Vietnam, along the insular region of south-east Asia (Nicobars, east through Sumatra, Borneo, Java, Sulawesi, Ceram, and Ambon) to the Philippines. The subspecies *kamaroma* ranges from north-eastern India, east to Indo-China, Peninsular Malaysia and Borneo.

BIOLOGY. This species is highly adaptable, and may be found living in ponds, mangrove swamps, freshwater marshes, canals, streams, and rice fields. It can also be found in plantations such as those of oil palm, rubber, orchards and in catchment areas in lowland forests and mangrove swamps, as well. It is equally common around human habitated environments where suitable habitats are available. Semi-aquatic, it can be encountered both on land and in water, although juveniles are more aquatic. It is a good swimmer and forages for aquatic plants, aquatic insects, molluscs and crustaceans. On land it feeds on plants, fungi and worms. Its favourite hiding place is within clumps of water weeds. This is a common pet turtle, especially in Europe and the USA, and a lot of information on its biology derive from observations made in captivity. Mating occurs in water, the female lays eggs on land. Females produce up to four clutches a season, each typically comprising 2–3 eggs, and exceptionally, 4–6 eggs. The elongated eggs are brittle and hardshelled, measuring 40–55 × 25–34 mm and weighing 15–32 grams. Incubation period is 70–100 days at 25–30°C, and hatchlings

---

(Opposite above). Malayan Box Turtle (*Cuora amboinensis*). (Below). Plastron. The transverse hinge across the plastron allows the shell to be closed off entirely, unlike in any other local turtle.

measure 38–48 mm in carapace length.

STATUS. The meat of this turtle is commonly eaten. It is the most common turtle sold in markets in Borneo and Peninsular Malaysia. Potential threats to the species include overexploitation for food and for the pet trade, pollution of waterways, and possibly, incidental capture in fishing gear. However, on account of its adaptability, its future is secure at present, as long as wetlands, including agricultural fields, exist. In the IUCN Red List of Threatened Animals the species is listed as 'Lower Risk: Near Threatened'.

REMARKS. The generic name *Cuora* applied to the Asian box turtles is derived from the Malay *Kura*, meaning hardshelled turtle.

Asian Leaf Turtle

Cyclemys dentata (Gray, 1831)
Iban *Beluku*; Malay *Bulus, Kura kura*

Emys dentata Gray, 1831. *Syn. Rept.* 1: 20.
Derivation: Latin for toothed, for the serrated marginals of juveniles.

IDENTIFICATION. The Asian Leaf Turtle has a wide, oval and depressed shell, rather leaf-like in shape. There are five vertebral scutes and the back of the forehead is divided into large shields. Carapace is tricarinate (although in very old individuals, the vertebral keel may not be distinct), depressed, more arched and elongated in adults and much flatter in juveniles. Plastron develops a hinge in turtles of carapace length 230–250 mm, making the plastral lobes of adults slightly moveable. Both the carapace and plastron are brown, sometimes with dark radiating lines.

It is a medium-sized turtle with a carapace length of 240 mm and weighs 1.5 kg. Shell colour ranges from brown or olive-brown to black, sometimes with black radiating lines. Plastron is yellowish-brown with dark radiating lines on each of the scutes, or uniformly dark brown or black. Juveniles usually have a chestnut tinge on the carapace and yellowish-brown venter, with each scute having dark

Bataguridae / Cyclemys dentata

Asian Leaf Turtle (*Cyclemys dentata*). Diagnostic characters include a depressed shell that typically bears three keels and plastron scutes with dark radiating lines.

Plastron of the Asian Leaf Turtle (*Cyclemys dentata*), showing the dark radiating lines in each scute.

radiating lines. Females outgrow males. Apart from this, sexual dimorphism is weak, tail size not different in the sexes, and males not showing plastral concavity.

DISTRIBUTION. The range of the species includes north-eastern India and possibly Nepal, east through Myanmar, Thailand, the Malay Peninsula, Sumatra, Java and Borneo. The *Cyclemys dentata* complex of turtles, as revised by Fritz *et al.* (1997) comprises four species, of which, in addition to *C. dentata*, one further species, *C. oldhamii* approaches the political borders of West Malaysia.

BIOLOGY. The preferred habitat of this species are small streams in hill forests and plains. Occasionally, individuals can be found in slow flowing larger streams especially at the base of small waterfalls as well as ponds. Juveniles are highly aquatic, while adults are more terrestrial, being essentially bottom-walkers rather than true swimmers. When picked up or threatened in any way, the turtle defaecates. These turtles are crepuscular, with activity restricted to dawn and at dusk. The diet of the species comprises both plant and animal (such as invertebrates, although carrion is also tackled) matter, and it is known to feast on figs. Nests are dug in the ground,

Asian Leaf Turtle (*Cyclemys dentata*). It has a wide, depressed shell, reminiscent of a dried leaf, and three keels on the carapace.

Bataguridae / Cyclemys dentata

and eggs are brittle hardshelled, measuring 57 × 35 mm, the incubation period being 82 days. Hatchlings are 56 mm in carapace length.

STATUS. This is still a common turtle within the overall range of the species, although hunted by indigenous people for food. Due to its drab colouration, it is not in great demand from the pet trade. *Cyclemys dentata* is listed under Action Plan Rating 3 of the IUCN/SSC Tortoise and Freshwater Turtle Specialist Group.

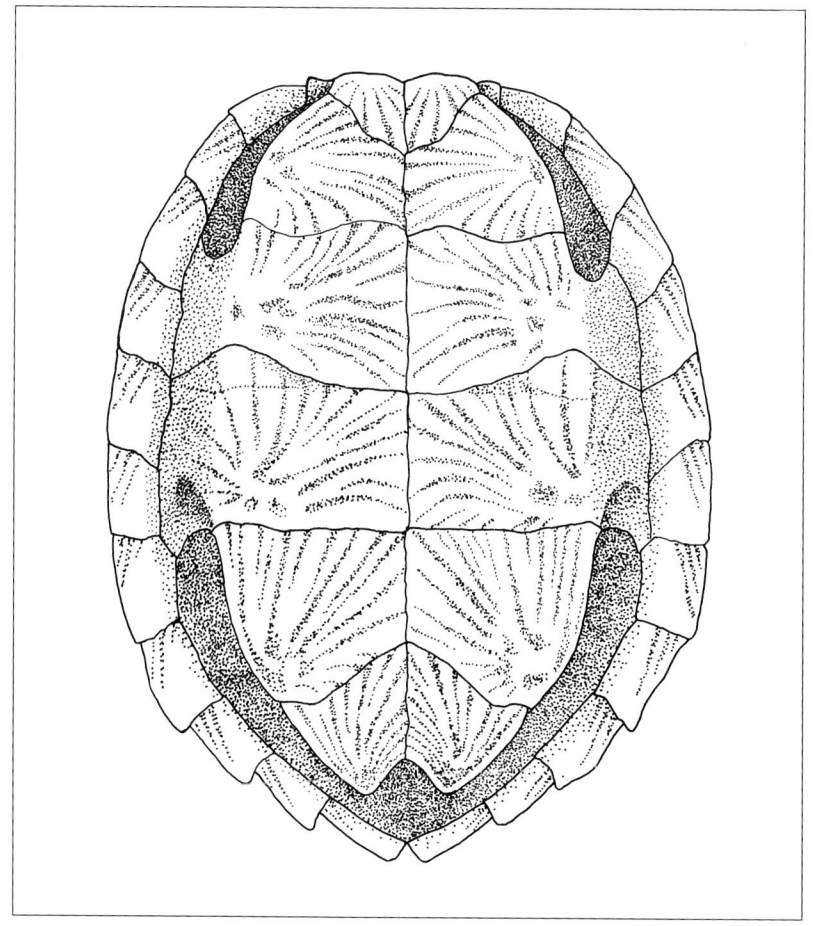

Plastron of the Asian Leaf Turtle (*Cyclemys dentata*) showing the radiating lines on the scutes.

Orange-headed Temple Turtle
Heosemys grandis (Gray, 1860)
Malay *Kura besar*

Geoemyda grandis Gray, 1860. *Ann. & Mag. nat. Hist. ser.* 6 3: 218. Derivation: Latin for large or big.

IDENTIFICATION. This is one of the largest freshwater turtles in the region, with a carapace length of up to 480 mm and weighing 12 kg. The shell is moderately elevated, vertebral keel distinct in adults; anterior marginals somewhat serrated, posterior marginals strongly serrated in adults. Hind lobe of plastron slightly movable in adult females (presumably to allow the passage of large eggs). Limbs relatively large, powerful, and protected by enlarged scales. Skin of back of forehead divided into large shields. Carapace dark brown to black, vertebral keel usually orange-brown. Plastron yellowish with fine dark brown lines radiating from areolus of each scute. Head orange with fine black specklings. Limb scales typically speckled with orange. Males outgrow females, in addition to showing a proportionally longer tail and a concave plastron.

Close-up of head of the Orange-headed Temple Turtle (*Heosemys grandis*).

(Above). Juvenile of Orange-headed Temple Turtle (*Heosemys grandis*). (Below). An adult: the vertebral keel is very distinct, and the marginals at the back of the shell are serrated.

DISTRIBUTION. The species ranges from southern Myanmar, through Thailand, Cambodia (possibly Laos) to southern Vietnam and Peninsular Malaysia. It has not been found in Singapore.

BIOLOGY. It is highly aquatic, inhabiting rivers, streams and marshes, and has been sighted crossing roads near streams. They are herbivorous in the wild, being known to eat figs, while captive animals may take animal matter. Clutches of 1–8 (4–6 from West Malaysia) elliptical eggs, measuring 65–67 mm, are laid. Hatchlings measure (carapace length) 42 mm and weigh 23 grams, emerging 99–113 days when the eggs are incubated at 28°C.

STATUS. Although its meat is eaten, more often captured turtles are given to temples to bring good luck to the donor. Although widespread, this species is not common. Nonetheless, large numbers of adults of this species are kept in local zoos and temple ponds. The IUCN Red List of Threatened Animals lists this species as 'Lower Risk: Near Threatened'.

Spiny Hill Turtle

Heosemys spinosa (Gray, 1831)
Malay *Kura kura, Kura kura duri bukit*

Emys spinosa Gray, 1831. *Syn. Rept.*: 20.
Derivation: Latin for spiny, for the shape of marginals in juveniles.

IDENTIFICATION. The Spiny Hill Turtle is easily recognised by the spiny serration on its carapace margin and pronounced vertebral keel. Juveniles in particular resemble pin cushions, with greatly expanded marginals having distinct spines that disappear or become less obvious with growth, or through wear and tear. It is thought that these spines act as a deterrent to predators, such as snakes. In adults, the serration is confined to anterior and posterior parts of the carapace and the vertebral keel, though still prominent, is slightly flattened. Posterior plastral lobe is kinetic in females, presumably to allow the passage of the large eggs. Back of thighs and base of tail have small conical scales; digits are partially webbed and hindlimbs club-shaped. Carapace reddish-brown in juveniles, brown in adults; plastron yellowish-brown, each plastral scute having dark radiating streaks. Limbs are greyish-brown; head brown, usually with a yellow spot behind eye. This is a medium sized turtle with a carapace length reaching up to 225 mm. Adult males possess relatively longer and thicker tails than adult females.

DISTRIBUTION. The range of the species includes southern Myanmar, through Thailand and Peninsular Malaysia and Singapore to the islands of Sumatra, Borneo and the Natunas (in Indonesia), to the Sulu Archipelago and Mindanao (in the Philippines).

BIOLOGY. This species is found both in lowland and hill forests. They have a fondness for wooded areas, with high humidity, and

Spiny Hill Turtle (*Heosemys spinosa*), with marginals that bear sharp points at the edges. The spiny marginals may deter enemies (such as snakes) from swallowing it, and with growth, the shell will become smoother.

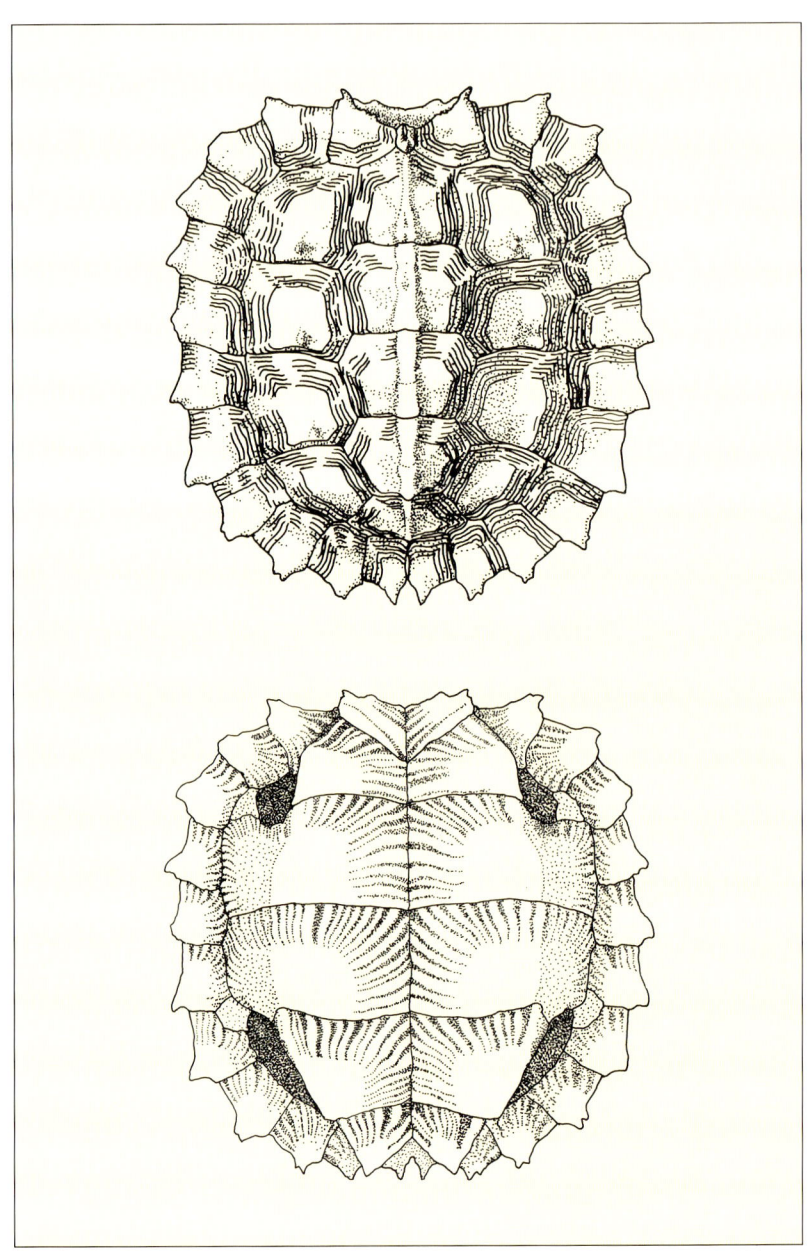

Carapace (above) and plastron (below) of the Spiny Hill Turtle (*Heosemys spinosa*).

Bataguridae / Heosemys spinosa

Plastron of a juvenile Spiny Hill Turtle (*Heosemys spinosa*), showing the attractive pattern of radiating lines and marginal spines.

generally stay under the leaf litter. Its shell colour blends well with that of dry leaves on the forest floor. A young turtle was once found in a shallow forest stream. These turtles are crepuscular, staying concealed during the heat of the day, and venturing out during the mornings and evenings. They are poor swimmers, but have a brisk walk. Although herbivorous in the wild, it feeds on both plants and animal matter in captivity, accepting food such as insects, worms, fish and meat on land. One to three elongated, brittle hardshelled eggs are laid, and to enable the passage of these relatively large eggs, a hinge develops in the plastron of the female, achieved by an incomplete separation of the abdominal scute into hyoplastral and hypoplastral elements and by fragmentation of an inguinal scute. Hatchlings measure 63.2 mm in carapace length. A young captive with a carapace length of 2 inches (= 51 mm), grew to 4.8 inches (= 102 mm) after three years and to 5.5 inches (= 139.7 mm) in five years, suggesting that growth is slow.

(Above). Plastron of an adult male Spiny Hill Turtle (*Heosemys spinosa*). (Below). An adult: note that the radiating lines on the plastral scutes and spiny marginals of adults are not as striking as in juveniles.

STATUS. The flesh of this species is consumed by the Orang Asli, the indigenous forest people of West Malaysia. The bizarre-looking juveniles are in demand as pets by zoos and private collectors. In Singapore, this species is restricted to the Central Catchment Area and the Bukit Timah Nature Reserve, and are sometimes sold as pets. The species is suspected to be threatened by deforestation and through collection for the pet trade and zoos. It is listed as 'Vulnerable' by the IUCN Red List of Threatened Animals.

Yellow-headed Temple Turtle
Hieremys annandalei (Boulenger, 1903)
Malay *Kura tokong*

Cyclemys annandalei Boulenger, 1903. *Fasc. Malayensis Zool.* 1: 142.
Derivation: For Thomas Nelson Annandale, founder of the Zoological Survey of India.

IDENTIFICATION. Carapace of juveniles raised, relatively more rounded, becoming elongated and flat-topped in adults. A vertebral keel, present in juveniles, disappear with growth. Posterior carapace margin serrated. Plastron truncate anteriorly, notched posteriorly. Skin of back of forehead smooth. Two large cusps on upper jaw, lower jaw with indentations at the corresponding site. Digits fully webbed. Carapace black, sometimes with an orange vertebral keel; plastron pale orange, occasionally with grey vermiculations, that eventually darken to become nearly a black plastron. Head is dark grey, with bright yellow lines along sides, extending to neck. Scales on limbs dark grey with yellow spots in adults, orange in hatchlings. A carapace length of 506 mm is attained. Mature males have a deep plastral concavity, longer, thicker tails, and are larger than adult females.

DISTRIBUTION. The species is known from Thailand, Cambodia, southern Vietnam and northern Peninsular Malaysia.

BIOLOGY. This turtle inhabits freshwaters, such as rivers, ponds, swamps and irrigation canals, and have also been recorded in estuarine areas. Almost entirely herbivorous, they feed on various species of macrophytes, such as leaves of the water lily (*Nymphaea* spp.). Captive specimens thrive on water plants, fruits and various types of vegetables. Mating takes places between November and March, with the nest being 5–10 m away from water, and 18–25 cm deep. A clutch typically contains 4–8 eggs, that are oblong, brittle-hardshelled, measuring 47–62 × 34–40 mm and 34–56 gm in weight. This species can be seen in temples of Perak and Penang. In Thailand large numbers are kept in the temple ponds in Bangkok. According to the tenets of Buddhism, a live gift offered to the temple gains merit for the donor in his next life. Unfortunately, the obligation ceases after the turtle's life has been saved, and as a result, hundreds of turtles are left to survive on tidbits offered by worshippers.

STATUS. The species is rare in West Malaysia, which may be due to past exploitation and habitat loss. It is listed as 'Vulnerable' by the IUCN Red List of Threatened Animals.

Yellow Headed Temple Turtle (*Hieremys annandalei*). The bright yellow facial patches are diagnostic of this turtle.

Malayan Snail-eating Turtle
Malayemys subtrijuga (Schlegel & Müller, 1844)
Malay *Jelebu siput*

Emys subtrijuga Schlegel & Müller, 1844. *In*: Temminck (ed.). *Verh. natuur. ges. Ned. Natuurk. Comm. Oost-Indie*: 30.
Derivation: Latin for nearly three-keeled; the species name could also imply a close relationship with an Indian freshwater turtle (*Melanochelys trijuga*).

IDENTIFICATION. The Malayan Snail-eating Turtle has a distinctive carapace, which has a prominent vertebral and two lateral keels. The plastron is large, its constituent bones immovable. The carapace is dark brown with yellow marginals. The plastron is yellow with a large black patch at the corner of each shield. Head is fairly

Malayan Snail-eating turtle (*Malayemys subtrijuga*). The three keeled carapace, large head and white or yellow facial stripes are distinguishing characters.

large, especially in old females, dark-brown with white or yellow stripes on the sides. The sides of the neck are marked with yellow streaks. Digits are entirely webbed. It reaches up to 200 mm. The tail is short and males have a proportionally narrower shell than females, besides a thicker tail base, a V-shaped anal notch (round in females), and smaller maximum size. Plastral concavity does not develop in males.

DISTRIBUTION. This small turtle occurs in the lowlands of Thailand, Cambodia, Laos and southern Vietnam, Kedah and Perlis in West Malaysia. It also occurs in Java and is thought to occur in Sumatra.

BIOLOGY. The main habitat of this species are the marshes but it can also be found in canals, ricefields and occasionally in slow flowing rivers, to areas with tidal influence. It has been recorded from the lower reaches of the Petani River. It is fairly common in southern Thailand, especially in ricefields and is locally known as the "ricefield turtle". In Peninsular Malaysia, it is not common and is

Close-up of head of the Malayan Snail-eating Turtle (*Malayemys subtrijuga*). The enlarged head is thought adaptive for a diet of snails.

restricted to the extreme north; this species does not occur in Singapore. It is active during the day and at night is more of a bottom-walker than a swimmer. Freshwater snails appear to be the primary food of juveniles turtles, as evidenced by stomach samples in a study conducted in Thailand that yielded snails of two species, *Filopaludina sumatrensis* and *Brotia costula*. Freshwater mussels are also eaten, as are shrimps, and perhaps crabs, insect larvae, worms and fishes. Nesting takes place between December and March in Thailand, and to April in southern Vietnam. Eggs are brittle-shelled and elliptical, measuring 37.9–42.4 mm, weighing 9.6–12.3 gm. Clutch sizes range between 3–6 eggs. Incubation periods for the species is 97–292 days. Hatchlings are around 35.3 mm in carapace length.

STATUS. There is hardly any information on the conservation status of this species in Peninsula Malaysia. The meat is eaten and it is quite possible that its rarity is the result of past exploitation for food. The species may be affected by urbanisation, and the use of agricultural pesticides. In Thailand, large numbers are sold for food; others are released into temple ponds and pools within city parks. The species is protected under Action Plan Rating 3 of the IUCN.

Malayan Flat-shelled Turtle

Notochelys platynota (Gray, 1834)
Malay *Kura punggung datar, Biuka*

Emys platynota Gray, 1834. *Proc. Zool. Soc. London* 1834: 54. Derivation: Latin for flat-backed.

IDENTIFICATION. The Malayan Flat-shelled Turtle can be recognised by its flat shell, with six or seven vertebral scales (all other freshwater turtles typically show five vertebral scales). The skin on the back of the forehead is divided into large shields with a low, interrupted vertebral keel. In juveniles, carapace almost circular, with anterior and posterior margins serrated, while in adults only the posterior margin is serrated. There is a weak plastral hinge; toes fully webbed. Carapace length reaches 360 mm. Adults have a more or less distinctive plastral hinge between the hyo- and hypoplastral bones, located under the pectoral-abdominal seam. Limbs are large and powerful, lower forelimbs with curved scales, and forelimbs with strong claws. Carapace rich brown with fine black pattern in adults, but hatchlings have black-spotted, green or bright yellow carapaces. Hatchling plastron orange, fading to yellow while a large black blotch develops on each scute that may grow to cover the entire scute, forming a black plastron. Head brown with large yellow or cream patches. Juveniles with longitudinal cream, yellow or orange lines, running from eye backwards to side of neck, another from jaw angle to throat. Markings most distinct in juveniles, and in adults become yellow on the snout, jaws and throat. Adult males, besides showing relatively longer and thicker tails, possess slightly concave plastra.

DISTRIBUTION. The range of the species includes southern Thailand, through Vietnam, West Malaysia and Singapore, to the islands of Sumatra, Java and Borneo.

Tan Heok Hui

A hatchling Malayan Flat-shelled Turtle (*Notochelys platynota*), showing the apple-green carapace. The carapace turns brown with growth.

Bataguridae / Notochelys platynota

(Above). Close-up of head of a Malayan Flat-shelled Turtle (*Notochelys platynota*). (Below). An adult: this is the only turtle in the region with six or seven vertebral scutes (all others have five).

BIOLOGY. This turtle can be found in shallow waters, such as marshes, swamps, ponds and streams in lowland rainforest situations, with an abundance of water plants. Largely herbivorous in the wild, it feeds primarily on water plants. In captivity it is omnivorous, accepting vegetables, fruits and animal matter including snails, earthworms, mealworms and crickets. They can feed both on land and in water. Two turtles of this species have been recorded attacking an Elephant Trunk Snake (*Acrochordus javanicus*). A 20.5 cm female produced three large, hardshelled eggs, measuring 56 × 27–28 mm. Hatchlings measure 55–56.2 mm in carapace length.

STATUS. A common, unprotected turtle, it is commonly caught for food by the local people and occasionally appear in the pet trade. The IUCN Red List of Threatened Animals considers this species as 'Data Deficient'.

Plastral view of a juvenile Malayan Flat-shelled Turtle (*Notochelys platynota*), showing the large dark areas.

Malayan Giant Turtle
Orlitia borneensis Gray, 1873
Malay (Malaya) *Juku juku besar;* Malay (Borneo) *Baning dayak*

Orlitia borneensis Gray, 1873. Ann. & Mag. nat. Hist. ser. 11 4: 157. Derivation: Latin for inhabitant of Borneo.

IDENTIFICATION. The Malayan Giant Turtle is the largest freshwater turtle on Borneo and in West Malaysia, growing to 800 mm in carapace length. It has a large head and a dark shell. The first vertebral scute is shaped like a mushroom. Its limbs are extensively webbed like in other large river turtles, and it lacks large scales on the

Malayan Giant Turtle (*Orlitia borneensis*), showing its large head and dark shell.

Turtles of Borneo and Peninsular Malaysia

Hans P. Hazebroek

Bataguridae / Orlitia borneensis

Malayan Giant Turtle (*Orlitia borneensis*) in its natural habitat in Batang Ai National Park, Sarawak, Borneo.

Turtles of Borneo and Peninsular Malaysia

(Above). Close-up of head of Malayan Giant Turtle (*Orlitia borneensis*). (Below). Plastron, showing the partially dark-pigmented plastron.

forelimbs. Denticulated edges of jaws are lacking, and skin back of forehead is divided into smaller shields. Carapace depressed, smooth, rather narrow, with four pairs of costals. In juveniles, it is convex and keeled posteriorly. Fingers and toes extensively webbed. Head dark, usually with a white spot below and behind angle of jaws. Adult males possess relatively longer and thicker tails. Adults are uniformly blackish-grey dorsally, with the underside yellowish to brownish-cream. Juveniles are brown above, yellow below.

DISTRIBUTION. This huge freshwater turtle is found in the large rivers and lakes of Peninsular Malaysia, Sumatra and Borneo.

BIOLOGY. This species lives in large rivers and estuaries and have been occasionally found in large, vegetation-chocked ponds. It is omnivorous, with a preference for fish. A large individual was observed capturing a Puff-faced Water Snake (*Homalopsis buccata*) in the vicinity of the Klang River. In captivity, it accepts vegetables, particularly Kangkong (*Ipomoea aquatica*) and soft fruits such as bananas and papayas. One in captivity accepted a variety of food including beef heart, horsemeat, canned dog food, and fish. It also

Malayan Giant Turtle (*Orlitia borneensis*). The adult female on left shows a relatively short tail and a wider postanal gap; the adult male on right shows a relatively longer tail and a shorter postanal gap.

accepted overripe bananas, but refused other vegetables. Food is taken as readily on land as in water. This species nests on river banks and river islands, females laying 12–15 eggs at a time, which are elongated and with brittle hard shells, measuring on average 80 × 40 mm. Hatchlings have shell lengths of about 60 mm and a rough texture to their shells with markedly serrated marginals at the back of the carapaces.

STATUS. This large river turtle is exploited for its meat. It was once abundant along the Klang River and is only occasionally seen at present. The IUCN Red List of Threatened Animals lists the species in the 'Lower Risk: Near Threatened' category.

Black Pond Turtle
Siebenrockiella crassicollis (Gray, 1831)
Malay *Kura kura kolam, Jelebu hitam*

Emys crassicollis Gray, 1831. *Syn. Rept.* 21.
Derivation: Latin for thick-necked.

IDENTIFICATION. The Black Pond Turtle is perhaps the second most common turtle in the region. Vertebral keel pronounced, juveniles with tricarinate carapaces, adults with unicarinate shell. Head large, skin of back of forehead divided into large scales. Carapace depressed and is black or dark brown, with yellow spots; plastron yellow with black spots or almost entirely black. In juveniles, large yellow facial spots are present. Limbs black and digits fully webbed. A carapace length of 200 mm is reached. Adult males with relatively longer and thicker tails, although slightly smaller than females. Additionally, they have a slightly concave plastra and lose the head spots, while females retain theirs.

(Opposite above). Black Pond Turtle (*Siebenrockiella crassicollis*). This is a very dark-coloured turtle, the carapace showing three keels in juveniles, but only one in adults. (Opposite below). Plastron.

Bataguridae / Siebenrockiella crassicollis

DISTRIBUTION. This species is distributed over southern Myanmar, Cambodia, south to Vietnam, Thailand, Peninsular Malaysia, Singapore, in addition to Sumatra, Java and Borneo.

BIOLOGY. This species inhabits deep ponds, marshes, sluggish streams around human habitated areas and water catchment areas in agriculture sites, even peat swamps, in lowland areas. It can walk on land and can be seen on roads in villages and between estates. They are bottom-walkers, rather than true swimmers, and spend a considerable amount of time submerged, rarely, if ever coming ashore, and when it does, it is generally at night. An omnivore, it feeds on plants, vertebrates and invertebrates (including worms, snails, shrimps and frogs), both on land and in the water, and is known to scavenge. When picked up, it exudes an odious secretion from its cloaca. In Peninsular Malaysia, 3–4 clutches are laid between April to June, each consisting of one or two eggs, although in captivity, clutch sizes of up to five are known. Eggs are elongated, brittle hardshelled, measuring 45 × 19 mm. Incubation periods of 60–80 days are known, and hatchlings are 52 mm in carapace length.

STATUS. The Black Pond Turtle is sometimes exploited for its meat and is sold in markets. The meat, however, is not always popular, due to its offensive odour.

Black Pond Turtle (*Siebenrockiella crassicollis*): close-up of head.

Chapter 7

EMYDIDAE
AMERICAN HARDSHELL TURTLES

A largely North American family, with 10 genera and 35 species. One species, the Red-eared Slider (*Trachemys scripta*), was introduced into West Malaysia and Singapore, and has established itself in many localities. These are hardshelled turtles, similar to the batagurids, and are primarily aquatic. The subspecies *scripta* was initially imported for the pet trade, but as they grow larger and are no longer colourful, their often teenage owners find them less attractive, and have apparently discarded large numbers of unwanted pets into the waterways of the region.

Red-eared Slider

Trachemys scripta (Schoepff, 1792)
Malay *Kura terlinga-merah*

Testudo scripta Schoepff, 1792. *Hist. Test. Icon. Ill*: 16.
Derivation: Latin for written. The name of the introduced subspecies, *elegans* is derived from the word meaning elegant.

IDENTIFICATION. The Red-eared Slider, a North American turtle, is now established in many parts of the tropics and subtropics,

Turtles of Borneo and Peninsular Malaysia

including West Malaysia and Singapore. It is easily recognizable by the bright head colouration and a lack of a plastral hinge. Shell rounded with an almost smooth margin and plastron lacking a hinge; carapace green with yellow lines to nearly black (old males often nearly black), plastron bright yellow with a large black ocellus on each scute. A distinct orange or red spot present behind eye. The brown head is yellow striped. A length of 280 mm is attained, with females outgrowing males. In addition, males possess very long claws on forelimbs.

Yong Hoi Sen

Red-eared Slider (*Trachemys scripta*): adult. The bright red temporal patch is unique to this exotic turtle introduced through the pet trade to many parts of Borneo and Peninsular Malaysia. This turtle is a native of North America.

Emydidae / Trachemys scripta

C.L. Chan

Plastron of an adult Red-eared Slider (*Trachemys scripta*).

DISTRIBUTION. While the species originates from North and Central America, this particular subspecies is from the Missisippi River drainage in the United States. It has become a pest in many temperate and tropical countries of the world, including Hawaii, Japan, Thailand, India, South Africa, France, Australia, Malaysia and Singapore.

BIOLOGY. Hatchlings were imported for the pet trade and sold by the thousands in local pet shops. Most owners have released adult turtles (that are less colourful) into the wild, and as a result, this species is now widespread in both rural and suburban areas in Peninsula Malaysia and Singapore. This is a cause for serious concern, as the Slider probably competes with the local turtle species and other components of the aquatic fauna. The turtle can be found in a wide variety of habitats, from temple ponds, to city ponds, lakes, small rivers, irrigation reservoirs and marshes. Omnivorous, its diet

I. Das

Basking adult Red-eared Slider (*Trachemys scripta elegans*). The head and limbs may be extended out of the shell as they bask at the edge of waterbodies.

Red-eared Slider (*Trachemys scripta*). (above). Hatchling. The bright carapace pattern gets obscured with growth. (below). Close-up of head, showing the bright red temporal patch, characteristic of the subspecies *elegans*.

C.L. Chan

Plastron of juvenile Red-eared Slider (*Trachemys scripta*).

consists of vegetation, fruit, small animals and even carrion. Clutches of 2–25 oval eggs, measuring 30–42 × 19–29 mm are laid, the with an incubation period between 65–75 days. Hatchlings measure 30–33 mm.

    STATUS. This turtle has been described as the most widely kept "pet" turtle in the world. The species is abundant in most areas where it has been introduced, and is obviously not of conservation concern. The IUCN Red List of Threatened Animals categorises the species as being at 'Lower Risk: Near Threatened'.

Chapter 8

TESTUDINIDAE
LAND TORTOISES

True land tortoises are grouped under a single family, the Testudinidae. It includes species that are entirely terrestrial, although a few may be fond of damp areas, such as edges of waterbodies. They have columnal forelimbs and club-shaped hindlimbs, and vaulted shells. Their digits are free, and the limbs bear large, overlapping scales. The head and limbs are entirely retractable within the shell. These are slow-moving, long-lived species, that are primarily herbivores, but may scavenge on dead animal matter. Three species occur in the region: the Elongated Tortoise (*Indotestudo elongata*), the Asian Brown Tortoise (*Manouria emys*), and the Impressed Tortoise (*M. impressa*).

Elongated Tortoise
Indotestudo elongata (Blyth, 1853)
Malay *Banding lontong, Kura kura mas*

Testudo elongata Blyth, 1853. *J. Asiat. Soc. Bengal* 22: 639.
Derivation: Latin for elongated, for its elongated shell.

IDENTIFICATION. The Elongated Tortoise is characterized by its high and elongated shell, with straight sides, a yellow carapace with black blotches (although in some animals, the shell is unpatterned), and a single, large supracaudal scute. The plastron is rigidly attached to the carapace by a long bridge. It reaches a carapace length of 360 mm. The anterior and posterior margins of the carapace are slightly reverted and strongly serrated in the young, feebly in aged specimens. The carapace is greenish-yellow or yellowish-brown, marked with a black spot or blotch in each costal and vertebral shield. The plastron is yellow with black markings in the center of every shield. Head is yellow, except during the breeding season, when adults have pink patches around their nostrils; the limbs are dark gray with yellow scales and its claws are either yellow or white. Males exceed females in size and have deep plastral concavities. They also show longer tail claws, comparatively narrower shells and suprapygals that recurve inwards. Females have a comparatively large post-anal gap, presumably to allow the passage of their large eggs.

Elongated Tortoise (*Indotestudo elongata*). It has a single, large supracaudal scute and the dark blotches on the carapace (shown in this photograph) are not always present.

Testudinidae / Indotestudo elongata

Plastron of the Elongated Tortoise (*Indotestudo elongata*).

DISTRIBUTION. This species ranges from northern India, from Uttar Pradesh, through Orissa and southern Nepal and eastern Bangladesh and north-eastern India, Myanmar through Thailand and West Malaysia.

BIOLOGY. They inhabit forested areas and in the Peninsula this species is confined to the northern regions. It is more common at higher elevations than in lowland forests. This species is relatively common in Thailand, where it is restricted to higher elevations. The species inhabits deciduous forests in the hills, at altitudes between 190–560 m. They are known to withstand heat up to 48°C, a remarkable capability for a tropical forest-dweller. This is achieved through salivation, wetting their head, neck and forelimbs, leading to evaporative cooling. With no clearly-defined activity periods, these tortoises may be most active in the early morning, late afternoon and early evening with activities peaking during and after a shower of rain. The Elongated Tortoises feed on fungi and slugs, leaves and shoots and fruits (*Ficus racemosa*, various *Dillenia* species, *Gmelina* *arborea* and *Shorea* sp.). Mushrooms are apparently a special favourite, and these animals are known to scavenge on animal kills. Adults establish dominance through intimidation of conspecifics by shell ramming. During the breeding season, the skin around the nostrils and eyes becomes pink, particularly in males. The timing of reproduction is apparently determined by the local environmental conditions, and generally coincides with the rains. Males intimidate females into submission by shell ramming. Eggs are laid in nest chambers underground, and clutches generally comprise one to five eggs, although a clutch size of nine eggs is on record. Up to two clutches per year can be laid by a single female. Eggs are cylindrical, with a brittle hardshell, measuring 50×37 mm, and take, depending on the locality, between 45–64 days to hatch, producing hatchlings with carapace lengths of 31–58 mm.

STATUS. This tortoise is occasionally found on sale in markets. The species is listed under Appendix II of CITES and Action Plan Rating 1 of the IUCN/SSC Tortoise and Freshwater Turtle Specialist Group.

Close-up of head of the Elongated Tortoise (*Indotestudo elongata*). During the breeding season, the face develops pink patches, especially in the males.

Asian Brown Tortoise

Manouria emys (Schlegel & Müller in: Temminck, 1844)
Malay (Peninsular Malaysia) *Baning perang, Kura kura anam kaki*
Kadazandusun (Sabah, Malaysian Borneo) *Suyan*

Testudo emys Schlegel & Müller, 1844. In: Temminck (ed.). *Verh.-natuur. ges. Ned. natuurk. Comm. Oost-Indie*: 34.
Derivation: From the Greek for turtle.

IDENTIFICATION. The Asian Brown Tortoise is the largest tortoises in Peninsula Malaysia with a carapace length of 560 mm and weighing up to 40 kg. The distinguishing features include a convex carapace flattened on top; annuli on vertebrals and costals prominent; supracaudals (or twelveth marginals) separate both dorsally and ventrally; upper jaw slightly hooked; a very large pointed tubercle on the base of thighs; a cluster of large scales on the thigh region; and tail terminating in a horny pointed scale. The carapace is dark brown

Asian Brown Tortoise (*Manouria emys*): close-up of head, showing the large forehead scales.

Turtles of Borneo and Peninsular Malaysia

Testudinidae / Manouria emys

(Left). Asian Brown Tortoise (*Manouria emys*). (Top). Close-up of scales on leg. (Bottom). Close-up of spur on hind leg, which is so well developed that the species is sometimes called the "six-footed tortoise".

Turtles of Borneo and Peninsular Malaysia

Asian Brown Tortoise (*Manouria emys*). (Top). Lateral view of shell. (Bottom left). Carapace. (Bottom right). Plastron. (From the collection of C.L. Chan).

to blackish-brown in adults, yellowish brown with dark brown markings in the young. The plastron is yellowish green to yellow in juveniles. The head and limbs are blackish-brown, being lighter in juveniles. Adult males may or may not develop a plastral concavity.

DISTRIBUTION. The distribution of the northern subspecies, *phayrei*, in south and the northern part of south-east Asia includes the mesic, broad-leaved forests of north-eastern India and south-eastern, and possibly also, north-eastern Bangladesh, Myanmar and Thailand. The southern subspecies (*emys*) is reported from southern China, Indo-China, southern Thailand, Peninsular Malaysia, and the islands of Sumatra and Borneo. The record from Java requires verification.

BIOLOGY. The Asian Brown Tortoise is a forest species, being recorded from rain forests and evergreen forests at elevations of up to 1000 m. They are fond of moisture, and are often found near streams, burrowing into mud, damp soil or piles of rotting leaves. A radio-

Asian Brown Tortoise (*Manouria emys*). The subcaudal (or the twelfth marginals) is divided both dorsally and ventrally and growth rings (annuli) on the shell are distinct.

Ventral view of the Asian Brown Tortoise (*Manouria emys*), showing the widely separated pectoral scutes that diagnose the subspecies *emys* from *phayrei*. This particular individual had a straight carapace length of 455 mm.

tagged female ranged in an area of 0.6 sq. km. Her locations were usually clustered for several days, followed by a walk of up to about 300 m. The species is largely herbivorous, eating bamboo shoots, tubers, mushrooms, seedlings and fallen figs, although invertebrates and frogs may also be consumed. Nests are constructed by the females by accumulating leaf-litter, soil and grass, adding leaf-litter by backsweeping with the forelimbs. On completion of egg laying, the nest is covered and packed using the hind limbs, following which the female guards the nest for 2–3 days. Eggs are laid in the months of April, May, September and October. Clutch sizes are in the range of 23–51. Eggs are white and nearly spherical with a diameter of 51–54 mm. Eggs take 63–69 days to hatch. Hatchlings measure 60–66 mm in carapace length.

STATUS. The species is favoured as food because of its large size (hence, more meat), and large number of empty shells of this species as be found in Orang Asli villages around forested areas of Peninsular Malaysia. Like the Yellow-headed Temple Turtle, the Asian Brown Tortoise is commonly seen in temples and occasionally, one comes across tortoises in the wild that has Buddhist inscriptions etched on their shells. Though the species is widely distributed throughout the Malay Peninsula, it is not abundant and some numbers are caught for export to the markets of East Asia. However, habitat destruction, including forest fires, must be the most serious threat to the survival of the species. The Asian Brown Tortoise is probably captured wherever it occurs throughout its range. The flesh is consumed and is believed to possess medicinal qualities, and in Bangladesh, the carapace is occasionally used as a feeding trough for chickens, pigs and dogs. Though the overall range of the species is large, its distribution is localized, and in many areas these tortoises have already become rare. The primary threats to the species appear to be exploitation for food and medicine, capture for zoos, as well as habitat destruction. The tortoise is listed as 'Insufficiently Known' in the IUCN Red Data Book, and is in Appendix I of CITES. It is also listed under Action Plan Rating 3 of the IUCN/SSC Tortoise and Freshwater Turtle Specialist Group and as 'Vulnerable' in the IUCN Red List. Protection of its forest habitats is required, as with all species that are linked to tropical moist forests.

Impressed Tortoise

Manouria impressa (Günther, 1882)
Malay *Banning bukit*

Geoemyda impressa Günther, 1882. *Proc. Zool. Soc. London* 1882: 343.
Derivation: Latin for pressed into, or impressed, for the flat-topped carapace.

Impressed Tortoise (*Manouria impressa*). The region around the vertebrals and costals are somewhat concave in adults and the scutes are translucent.

Testudinidae / Manouria impressa

Plastron of the Impressed Tortoise (*Manouria impressa*).

IDENTIFICATION. The Impressed Tortoise is characterized by a flattened shell and strikingly beautiful shell colouration. Carapace flattened on the vertebral region, with vertebral and costal region slightly concave in adults, flattened in juveniles. Anterior and posterior margins strongly serrated as well as reverted. A pair of small supracaudal scales are present over the tail. Forearm and feet scales raised and pointed. A single pronounced conical spur present on each thigh. A carapace length of up to 302 mm and weight to 3.7 kg are attained. Carapace scutes translucent, brown or orange-yellow, streaked with brown, black or orange. Plastron yellowish-brown, with more or less distinct dark radiating streaks. Head is yellow, limbs brown. Juveniles are light yellowish brown, finely speckled with black dots. Males have a relatively larger tail.

DISTRIBUTION. Although the Impressed Tortoise has a wide distribution, it appears to be highly localised. The overall range extends from northern and western Myanmar, Thailand, West Malaysia, Laos, northern Vietnam, Yunnan and Hainan (in China).

BIOLOGY. This highly specialised tortoise is confined to evergreen and bamboo forest, between 900–1200 m. In captivity, its primary food is mushrooms; diet in the wild being unknown. In Thailand, mating occurs between mid-March and September. A shallow nest is excavated, and 17–20 eggs may be laid, measuring 44 × 40 mm and weighing 39 gm apiece.

STATUS. There are few records of this species in Peninsula Malaysia, including Frazer's and Maxwell's Hills and the Cameron Highlands in Pahang. The apparent rarity of the species throughout its range may be a reflection of its specialised habitat requirements—primary forests at mid-altitudes. Therefore, habitat destruction must be a significant threat to the survival of the species. As is the fate of all other medium or large tortoises in the world, the present species may additionally be vulnerable to hunting by humans in the remaining pockets where it occurs. This tortoise is in high demand from the pet trade, although mortality in captivity is high. It is protected under Appendix II of CITES and listed in the IUCN Red List of Threatened Animals as 'Vulnerable'.

Close-up of head of the Impressed Tortoise (*Manouria impressa*).

Chapter 9

CONSERVATION

The belief that many wild species have aphrodisiac and various other medicinal properties, along with the demand for wild meat as connoisseur fare are important reasons for the endangerment of many species in south-east Asia. This is particularly true of turtles as well, and the Malayan Box Turtle (*Cuora amboinensis*) is by far the most commonly sold species in markets of Sabah, Sarawak and Peninsular Malaysia. In these markets and at restaurants, it is also possible to see softshell turtles, particularly the Asian Softshell Turtle (*Amyda cartilaginea*) and sometimes also the Malayan Softshell Turtle (*Dogania subplana*), kept in aquaria on display. The supply for turtles at present to restaurants in Peninsular Malaysia from Thailand, Burma, and even further afield, such as India, is abundant proof that there has been a serious depletion of wild turtle stocks locally.

Eggs of sea turtles are frequently collected for consumption—either on spot or are harvested for sale in local markets. Over-harvesting of eggs have been the reason for the endangerment of all sea turtles, as well as two hardshell turtles, the River Terrapin (*Batagur baska*) and the Painted Terrapin (*Callagur borneoensis*), all of which lay eggs in coastal areas where egg collection used to be an organised industry.

Turtles are also caught for the pet trade and for zoos, the demand being mostly from the West, and profit margins being large, large numbers were at one time harvested from the wild for European and American markets. The indigenous people of the forests of Borneo and Peninsular Malaysia have traditionally hunted turtles, and their impact on turtle populations will presumably remain insignificant as long as sophisticated fishing gear (such as harpoon guns, boats with outboard motors and other imports from the outside world) are not used.

Coupled with obvious threats such as harvest for the market is that of deforestation. Both Borneo and Peninsular Malaysia have seen a significant decrease in their forested land in the last 50 years, and the impact of logging (for the valuable hardwoods, an important source of foreign exchange) on forest-dwelling species of turtles must have been significant. Logging trails also make areas once inaccessible open to further intrusion, and this, in turn, has lead to once remote localities within reach of collectors. Industrialisation and urbanisation are other factors that have affected turtles, the opening up of land for these activities surely not beneficial for forest-dwelling species that are intolerant of open or polluted areas. Finally, forest fires, such as those currently raging over Borneo, are an obvious threat to not only ground-dwelling turtles, that are generally too slow

A *Baning dayak* (*Orlitia borneensis*) that was caught by hook and line near Lake Gempang, Kalimantan, Indonesian Borneo.

to avoid being burnt, but also many other forest-adapted plants and animals.

Perhaps because of an aquatic mode of life for a majority of species, turtles are covered by legislation for fisheries rather than wildlife. Consequently, their conservation and management remain in the hands of the Fisheries Department (which is primarily concerned with revenue earning activities, rather than conservation), instead of the Department of Wildlife and National Parks. This apparent anomaly needs to be corrected to enhance protection for the local turtle species, and new legislation, based on the conservation status of individual turtle species, needs to be drawn up and implemented.

In summary, turtles do not provoke the same strong sense of fear, hatred or dislike in people, as do most other reptile groups. In fact, many species are hugely popular, and in demand from the pet trade and from zoos, which now, along with deforestation and other forms of habitat change, threaten many of the local species. Others are threatened by the overharvest of eggs, and collection for food or medicine.

Turtles play a number of important functions in the ecosystems, such as controlling invertebrate pests and invasive weeds, providing food for other animals, and scavenging on dead and injured animals, thereby helping in the release of locked-up nutrients back to the environment. Unfortunately, too little is known of the biology of the local turtle fauna. We still need to answer many basic questions on most of our species, such as where they are found, what they eat and when or how often they breed. However, information derived from research itself will be useful only when put to use by planners and law makers in charge of the wilderness areas of Borneo and Peninsular Malaysia.

Acknowledgements

LBL wishes to acknowledge the help given by the pioneering Malaysian Zoologist, the late Prof. John L. Harrison who guided him through the basics of zoology, and to the late Prof. J.R. Audy, who was a special mentor in later studies.

A large number of our colleagues aided our work on tropical herpetofauna over the years. These include (in alphabetical order): Jasmi Abdul, Kraig Adler, Harry V. Andrews, Aaron Bauer, James Buskirk, Joseph Charles, Patrick David, Stuart Davies, Anslem De Silva, Bernard Devaux, David Edwards, Balazs Farkas, John G. Frazier, Uwe Fritz, Maren Gaulke, Richard Gemel, Robert F. Inger, John B. Iverson, Kiew Bong Heang, Kelvin K.P. Lim, Lim Ching Chiz, William P. McCord, Edward O. Moll, Illar Muul, Jarujin Nabhitabhata, Peter Ng, Wirot Nutaphand, Pang Khang Aun, Rohan Pethiyagoda, Peter C.H. Pritchard, Louis Ratnam, Dionysius Sharma, Nipon Srinarumol, Kumthorn Thirakhupt, Peter Paul van Dijk, Van Wallach, Robert G. Webb, Romulus Whitaker, M. Nordin, Norsham Yaakob, Yong Hoi Sen and Er-Mi Zhao. Our researches received the generous support of the British Council, a Fulbright Fellowship, the Centre for Herpetology, Madras, Universiti Brunei Darussalam and Universiti Malaysia Sarawak.

Curatorial staff of the following institutions provided facilities and access to specimens under their care: American Museum of Natural History (AMNH), New York (Darrel Frost and Linda Ford); The Natural History Museum (BMNH), London (E. Nicholas Arnold and Colin McCarthy); Field Museum of Natural History (FMNH), Chicago (Harold K. Voris, Robert F. Inger and Alan Resetar); Museum of Comparative Zoology (MCZ), Cambridge, MA (John Cadle, José Rosado and Ernst E. Williams); Muséum National d'Histoire Naturelle (MNHN), Paris (Roger Bour, Alain Dubois, Ivan

Ineich); Naturhistorisches Museum, Wien (NMW), Vienna (Richard Gemel, Heinz Grillitsch, Franz Tiedermann); Oxford University Zoological Museum (OUM), Oxford (Tom Kemp); Sarawak Museum (SM), Kuching (Peter Kedit and Charles Leh); Sabah State Museum (SSM), Kota Kinabalu (Joseph Pounis Guntavid and Anna Wong); Universiti Malaysia Sabah (USM), Kota Kinabalu (Maryati Mohammed, Axel Poulsen and Lucy Kimsui); Florida Museum of Natural History (UF), Gainesville (David Auth and F. Wayne King); the Raffles Museum of Biodiversity Research, National University of Singapore (USDZ/ZRC), Singapore (Kelvin Lim, Peter Ng and C.M. Yang); the National Museum of Natural History, Smithsonian Institution (USNM), Washington, D.C. (Ronald I. Crombie, Ronald W. Heyer and George R. Zug); and Zoological Survey of India (ZSI), Calcutta (J.R.B. Alfred, S.K. Chanda and N. C. Gayen).

The following colleagues supplied turtles for photography: Chai Kok Shin, Wong Ting Sung, Postar Muin, Kittipong Jarutanin, and William P. McCord. Additional photos for the book were supplied by C.L. Chan, Arthur Y.C. Chung, Christian Gönner, Hans P. Hazebroek, Jason Isley of Scubazoo Video Sdn. Bhd., Ulrich Manthey, Nicholas Pilcher, W.M. Poon, Cede Prudente, Francis Seow-Choen, Dionysius Sharma, Tan Heok Hui, and Yong Hoi Sen. Paper Land Sdn. Bhd. provided permission for use of their photograph of the Leatherback Sea Turtle.

For page setting and general assistance, we thank Cheng Jen Wai, Chan Hin Ching, Chan Shuk Ling, Chan Shuk Mun, Yong Ket Hyun, Jacqueline Chu, and Wong Mui Yun. Wong Khoon Meng provided valued comments on the manuscript.

We thank Y.B. Datuk Chong Kah Kiat, Minister of Tourism Development, Environment, Science and Technology, Sabah for presenting the foreword and Mr Joseph Guntavid, Director of Sabah Museum for writing the preface. Finally, ID wishes to thank Universiti Malaysia Sarawak and Prof. Ghazally Ismail for institutional support during the preparation of this manuscript. Last but not the least, we would like to thank Chan Chew Lun, our publisher, for inviting us to write this book, and for being positive during its production.

Glossary

adult: sexually mature individual.

anterior: nearer the front (towards the head).

arribada: aggregation of the Ridley Sea Turtles on nesting beaches.

aquatic: species that live in water.

beak: sheath-like covering of the jaws, comprising a single horny plate over each jaw surface.

carapace: the dorsal shell of a turtle (see p. 10).

circumtropical: encircling the earth, between 23°30' N and 23°30' S.

CITES: acronym for the Convention on International Trade in Endangered Species of Wild Fauna and Flora (also called the Washington Convention), which provides protection to wild species threatened by international trade.

clutch: the total number of eggs laid by a female at a time.

clutch size: number of eggs in a nest.

courtship: behaviour preceding mating.

crepuscular: active during dawn and dusk.

depressed: flattened from top to bottom.

dermal: pertaining to the skin.

dimorphic: the existence of two different forms (e.g., male and female) in the same species.

disc: the dorsal or ventral shell, especially of a Softshell Turtle.

dorsum: (as used here) the dorsal surface of the body, excluding head and tail.

Glossary

freshwater: water which is not salty, for example pond or river water.

imbricate: overlap, as the scales on the carapace of Hawksbill Sea Turtles.

inframarginal pore: opening located near the end of the inframarginal scutes in the Ridley Sea Turtles that give out secretions.

inguinal: a scute in the posterior region of the plastron.

juvenile: a sexually immature individual.

keel: a narrow prominent ridge.

mandible: lower jaw.

marginals: the series of scutes on the outer margin of the carapace.

marine: pertaining to the sea.

migration: periodic movement of animals from one place to another.

nocturnal: active during the night.

nuchal: anterior-most scute of the carapace, close to the neck.

parietal: paired scales on the forehead, behind the frontals, forming the last pair of large scales on the head.

plastron: the ventral shell of a turtle (see p. 10).

prefrontals: paired scales on the anterior margin of the orbit of the eye, usually bounded by the frontal.

proboscis: the external fleshy part of the nose of a Softshell Turtle.

pyrosoma: a group of tunicates that line in colonies, embedded in the walls of a gelatinous tuber. They have bright bioluminescence.

salinity: measure of salt concentration of water. Higher salinity implies more dissolved salts.

scute: a horny epidermal shield.

serrated: possessing a saw-toothed edge.

supracaudal: the posterior-most scute of the carapace in turtles and tortoises.

suprapygal: the posterior-most bony element of the carapace that is situated over the vertebra.

tortoiseshell: jewellery manufactured from scutes of sea turtles. The best tortoiseshell comes from the Hawksbill Sea Turtle, although it may also be produced from the Green Turtle.

tubercle: a knot-like projection.

ventrum: (as used here) the ventral surface of the body, excluding head and tail.

vertebral: pertaining to the region of the backbone. Also the central series of scutes on the carapace of turtles.

References

Aikanathan, S. (1996). Malaysian Sea Turtles: what Future? *Marine Turtle Newsletter* (73): 23–24.

Arumugam, S.R. (`976). The River Terrapin or Tuntong (*Batagur baska*). *Malayan Naturalist* 3(1 & 2): 8-10.

Azrina, L.A. and B.L. Lim (1999). Legislative Status of Chelonian species in Selangor, Peninsular Malaysia. *Malayan Nature Journal* 53(3): 253–261.

Balasingam, E. (1965). Conservation of Green Turtles (*Chelonia mydas*) in Malaya. *Malayan Nature Journal* 19: 235–236.

Balasingam, E. (1965).The Giant Leathery Turtle Conservation Programme, 1964. *Malayan Nature Journal* 19: 145-146.

Balasingam, E. (1967). The Ecology and Conservation of the Leathery Turtle, *Dermochelys coriacea* (Linn.) in Malaya. *Micronesia* 3: 37–43.

Balasingam, E. (1965).Turtle Conservation Results of 1965 Hatchery Programme. *Malayan Nature Journal* 20: 139-141.

Balasingam, E. and M. Khan bin M. Khan (1969). Conservation of the Perak River Terrapin *(Batagur baska)*. *Malayan Nature Journal* 23: 27–29.

Balasingam, E. and Y.P. Tho (1972). Preliminary Observation on Nesting returns of the Leathery Turtle (*Dermochelys coriacea*) in Central Trengganu, Malaysia. *Malayan Nature Journal* 25: 5–9.

Banks, E. (1937). The Breeding of the Edible Turtle (*Chelonia mydas*). *Sarawak Museum Journal* 4: 523–532.

Bartlett, D. and P. Bartlett (1999). The Pond Turtles: Pretty as ever and still going strong. *Reptiles* 7(7): 48–59.

Bhaduria, R.S., A. Pai and D. Basu (1990). Habitat, Nesting and Reproductive Adaptations in Narrow-headed Softshell Turtle *Chitra indica* (Gray), (Reptilia: Chelonia). *Journal of the Bombay Natural History Society* 87(3): 364–367.

Biswas, S. (1981). A Report on the Olive Ridley, *Lepidochelys olivacea* (Testudines: Cheloniidae) of Bay of Bengal. *Records of the Zoological Survey of India* 79: 275–302.

Biswas, S., L.N. Acharjyo and S. Mohapatra (1978). Notes on Distribution, Sexual Dimorphism and Growth in Captivity of *Geochelone elongata* (Blyth). *Journal of the Bombay Natural History Society* 75: 928–930.

Bjorndal, K.A. (ed.) (1982). *Biology and Conservation of Sea Turtles.* Smithsonian Institution Press, Washington, D.C. (2) + 615 pp. (Revised edition, 1996), with a new preface.

Bonin, F., B. Devaux and A. Dupré (1996). *Toutes les Tortues du Monde.* Delachaux et Niestlé, Lausanne, Switzerland. (2) + 254 pp.

Boulenger, G.A. (1903). *Fasciculi Malayenses. Anthropological and Zoological Results of an Expedition to Perak and the Siamese Malay States, 1901–1902. Zoology. Pt. I. Report on the Batrachians and Reptiles.* Pp. 131–176; 5 pl.

Boulenger, G.A. (1912). *A Vertebrate Fauna of the Malay Peninsula from the Isthmus of Kra to Singapore including the Adjacent Islands. Reptilia and Batrachia.* Taylor & Francis, London. xiii + 294 pp.

References

Bourret, R. (1941). *Les Tortues de l'Indochine.* Institut Océanographique de l'Indochine, Hanoi. 235 pp.

Bramble, D. (1974). Emydid Shell Kinesis: Biomechanics and Evolution. *Copeia* 1974: 707–727.

Buskirk, J.R. (1997). The Malayan Flat-shelled Turtle, *Notochelys platynota. The Vivarium* 9(1): 6–9, 15.

Bustard, H.R. (1972). *Sea Turtles. Natural History and Conservation.* William Collins Sons and Co. Ltd., London. 220 pp.

Cantor, T.E. (1847). Catalogue of Reptiles inhabiting the Malay Peninsula and Islands. *Journal of the Asiatic Society of Bengal* 16: 607–656.

Chan, E.H. and H.C. Liew (1989). *The Leatherback Turtle. A Malaysian Heritage.* Tropical Press Sdn. Bhd., Kuala Lumpur. vii + 49 pp.

Chan-ard, T., K. Thirakhupt and P.P. van Dijk (1996). Observations on *Manouria impressa* at Phu Luang Wildlife Sanctuary, Northeastern Thailand. *Chelonian Conservation and Biology* 2 (1): 109–113.

Chin, L. (1975). Notes on Marine Turtles (*Chelonia mydas*). *Sarawak Museum Journal* 23: 259–265.

Chou, L.M. and B.L. Choo (1986). The Potential of Soft-shell Turtle Culture in Singapore. In: *Proceedings of the International Conference on Development and Management of Living Aquatic Resources, Serdang, Malaysia, August 1983.* Pp. 177–180.

Christiansen, J. (1973). *Heosemys spinosa. Nordisk Herpetologisk Forening* 16(5): 90–95. [In Danish.]

Chua, T.H. (1979). Conservation of the Hawksbill Turtle in Malacca. Can it be a Reality? *Malayan Nature Journal* 33(2): 15–17.

Cox, M.J., P.P. van Dijk, J. Nabhitabhata and K. Thirakhupt (1998). *A Photographic Guide to Snakes and other Reptiles of Peninsular Malaysia, Singapore and Thailand.* New Holland Publishers (UK), Ltd., London. 144 pp.

Crumly, C.R. (1982). A Cladistic Analysis of *Geochelone* using Cranial Osteology. *Journal of Herpetology* 16(3): 215–234.

Crumly, C.R. (1984). A Hypothesis for the Relationship of Land Tortoise Genera (family Testudinidae). Studia Geologica Salmanticensia. Vol. Especial I. *Studia Palaeochelonologica* 1: 115–124.

Das, I. (1986). The Diversity and Utilisation of Land Tortoises in Tropical Asia. *Tigerpaper* 13(3): 18–21.

Das, I. (1989). Asian Land Tortoises. *Sanctuary Asia* 9(4): 40–47.

Das, I. (1991). *Colour Guide to the Turtles and Tortoises of the Indian Subcontinent.* R & A Publishing Limited, Portishead. iv + 133 pp + 16 pl.

Das, I. (1993). Vernacular Names of some Southeast Asian Amphibians and Reptiles. *Sarawak Museum Journal* 34: 123–139.

Das, I. (1995). *Turtles and Tortoises of India.* Oxford University Press, Bombay. x + 176 pp + 16 pl.

Das, I. (1995). An Illustrated Key to the Turtles of Insular South-east Asia. *Hamadryad* 20: 27–32.

David, P. (1994). Liste des Reptiles Actuels du Monde I. Chelonii. *Dumerilia* 1: 7–127.

De Rooij, N. (1915). *The Reptiles of the Indo-Australian Archipelago. I. Lacertilia, Chelonia, Emydosauria.* E.J. Brill, Leiden. xiv + 384 pp.

References

De Silva, G.S. (1969). Turtle Conservation in Sabah. *Sabah Society Journal* 5: 6–26.

De Silva, G.S. (1969). Marine Turtle Conservation in Sabah. *Annual Report for Research Board, Forest Department* 1969: 124–135.

De Silva, G.S. (1971). Marine Turtles in the State of Sabah, Malaysia. *IUCN n.s. Supplement Paper* 31: 47–52.

De Silva, G.S. (1978). Turtle Notes (Sightings of *Dermochelys* in Sabah waters). *Borneo Research Bulletin* 10(1): 23–24.

De Silva, G.S. (1980). The Status of Sea Turtle Populations in East Malaysia and the South China Sea. In: K.A. Bjorndal (ed.). *Biology and Conservation of Sea Turtles.* Pp. 327–337. Smithsonian Institution Press, Washington, D.C.

Diong, C.H. (1976). Case for a Sanctuary on Pangkor Island. *Malayan Naturalist* 2(3 & 4): 8–9.

Dodd, C.K. (1988). Synopsis of the Biological Data on the Loggerhead Sea turtle *Caretta caretta* (Linnaeus 1758). *U.S. Fish and Wildlife Service Biological Report* 88(14): 1–110.

Ernst, C.H., R.W. Barbour and R. Altenberg (1998). *Turtles of the World.* CD-ROM version. Springer Verlag, Berlin.

Ernst, C.H., R.W. Barbour and J.E. Lovich (1994). *Turtles of the United States and Canada.* Smithsonian Institution Press, Washington, D. C. xxxviii + 578 pp.

Ewert, M.A. (1979). The Embryo and the Egg: Development and Natural History. In: M. Harless & H. Morlock (eds.). *Turtles: Perspectives and Research.* Pp. 333–416. Wiley Interscience, New York.

Ewert, M.A. (1985). Embryology of Turtles. In: C. Gans (ed.). *Biology of the Reptilia*. Vol. 14. Development. Pp. 75–267. John Wiley & Sons, Inc., New York.

Farkas, B. (1994). Notes on Type and Type Locality of the Narrow-headed Softshell Turtle, *Chitra indica* (Gray, 1831) (Testudines, Trionychidae). *Miscellanea Zoologica Hungarica* 9: 117–119.

Frazier, J.G. and I. Das (1994). Some Noteworthy Records of Testudines from the Indian and Burmese Subregions. *Hamadryad* 19: 47–66.

Gibbons, J.W. and R.D. Semlitsch (1982). Survivorship and Longevity of a Long-lived Vertebrate species: How Long Do Turtles Live? *Journal of Animal Ecology* 51: 523–527.

Greer, A.E., J.D. Lazell and R.M. Wright (1973). Anatomical Evidence for a Counter-current Heat Exchanger in the Leatherback Turtle (*Dermochelys coriacea*). *Nature* 244: 181.

Groombridge, B. and L. Wright (compiled) (1982). *The IUCN Amphibia-Reptilia Red Data Book. Part 1- Testudines Crocodylia Rhynchocephalia*. IUCN, Gland. (4) + xiii + 426 pp.

Grossmann, W. (1986). *Callagur borneensis* (Schlegel & Müller). *Sauria, Berlin* 8(4): 1–2.

Harrisson, T. (1951). The Edible Turtle (*Chelonia mydas*) in Borneo: 1. Breeding Season. *Sarawak Museum Journal* 3: 593–596.

Harrisson, T. (1954). The Edible Turtle (*Chelonia mydas*) in Borneo: 2. Copulation. *Sarawak Museum Journal* 6: 126–128.

Harrisson, T. (1955). The Edible Turtle (*Chelonia mydas*) in Borneo: 3. Young Turtles (in captivity). *Sarawak Museum Journal* 6: 633–640.

Harrisson, T. (1956). The Edible Turtle (*Chelonia mydas*) in Borneo: 4. Growing Turtles and Growing Problems. *Sarawak Museum Journal* 7: 233–239.

Harrisson, T. (1956). The Edible Turtle (*Chelonia mydas*) in Borneo: 5. Tagging Turtles (and Why). *Sarawak Museum Journal* 7: 504–515.

Harrisson, T. (1958). Notes on the Edible Turtle (*Chelonia mydas*): 6. Semah Ceremonies, 1949–1958. *Sarawak Museum Journal* 8: 482–486.

Harrisson, T. (1958). Notes on the Edible Turtle (*Chelonia mydas*): 7. Long-term Tagging Returns, 1952–1958. *Sarawak Museum Journal* 8: 772–774.

Harrisson, T. (1959). Notes on the Edible Turtle (*Chelonia mydas*): 8. First Tag Returns outside Sarawak. *Sarawak Museum Journal* 9: 277–278.

Harrisson, T. (1961). Notes on the Green Turtle (*Chelonia mydas*): 9. Some New Hatching Observations. *Sarawak Museum Journal* 10: 293–299.

Harrisson, T. (1962). Notes on the Green Turtle (*Chelonia mydas*): 10. Some Emergence Variations. 11. West Borneo Numbers, the Downward Trend. 12: Monthly Laying Cycles. *Sarawak Museum Journal* 10: 610–630.

Harrisson, T. (1963). Notes on Marine Turtles: 13. Growth Rate of the Hawksbill. 14: Albino Green Turtles and Sacred Ones. *Sarawak Museum Journal* 11: 302–306.

Harrisson, T. (1964). Notes on Marine Turtles- 15: Sabah's Turtle Islands. *Sarawak Museum Journal* 11: 624–627.

Harrisson, T. (1965). Notes on Marine Turtles- 16: Some Loggerhead and Hawksbill Comparisons with the Green Turtle. *Sarawak Museum Journal* 12: 419–422.

Harrisson, T. (1966). Notes on Marine Turtles- 17: Sabah and Sarawak Islands Compared. *Sarawak Museum Journal* 14: 335–340.

Harrisson, T. (1967). Notes on Marine Turtles- 18: A Report on the Sarawak Turtle Industry (1966) with Recommendations for the Future. *Sarawak Museum Journal* 15: 424–436.

Harrisson, T. (1969). The Marine Turtle Situation in Sarawak. *IUCN Publ. (Occ. Pap.)* 20: 171–173.

Hendrickson, J.R. (1958). The Green Turtle, *Chelonia mydas* (Linn.) in Malaya and Sarawak. *Proceedings of the Zoological Society, London* 130: 455–533.

Hendrickson, J.R. and E.R. Alfred (1961). Nesting Populations of Sea Turtles on the East Coast of Malaya. *Bulletin of the Raffles Museum* 26: 190–196.

Hirayama, R. (1984). Cladistic Analysis of Batagurine Turtles (Batagurinae: Emydidae: Testudinoidea): A Preliminary Result. Studia Geologica Salmanticensia. Vol. Especial 1. *Studia Palaeocheloniologica* 1: 141–157.

Hirth, H.F. (1971). Synopsis of Biological Data on the Green Turtle, *Chelonia mydas* (Linnaeus) 1758. *FAO Fisheries Synopsis* 85: 1–8, 19.

Hoogmoed, M.S. and C.R. Crumly (1984). Land Tortoise Types in the Rijksmuseum van Natuurlijke Histoire with Comments on Nomenclature and Systematics (Reptilia: Testudines: Testudinidae). *Zoologische Mededelingen* 58(1): 241–259.

IUCN. (1996). *1996 IUCN Red List of Threatened Animals.* IUCN, Gland. 448 pp.

IUCN/SSC Tortoise and Freshwater Turtle Specialist Group. (1989). *Tortoises and Freshwater Turtles—an Action Plan for their Conservation.* IUCN, Gland. 41 pp.

Iverson, J.B. (1992). *A Revised Checklist with Distribution Maps of the Turtles of the World.* Privately printed, Richmond, Indiana. xiii + 363 pp.

Jasmi, A. (1986). Experimental Study of Hatching Softshell Turtle *(Trionyx cartilagineus)* Eggs Artificially at Kuala Tahan, Taman Negara. *Journal of Wildlife and Parks* 5: 119–124.

Jenkins, M.D. (1995). *Species in Danger. Tortoises and Freshwater Turtles: The Trade in Southeast Asia.* Traffic International, Cambridge. iv + 48 pp.

Khan, M. bin M. Khan (1964). A Note on *Batagur baska* (the River Terrapin or Tuntong). *Malayan Nature Journal* 18: 184–187.

Kiew, B.H. (1984). Conservation Status of the Malaysian fauna. IV. Turtles, Terrapins and Tortoises. *Malayan Naturalist* 38(2): 2–3.

King, F.W. and R.L. Burke (1989). *Crocodilian, Tuatara, and Turtle Species of the World: A Taxonomic and Geographic Reference.* Association of Systematics Collections, Washington, D. C. xxii + 216 pp.

Kitana, N. (1998). *Sexual Dimorphism and Annual Reproductive Cycle of the Common Asiatic Softshell Turtle* Amyda cartilaginea. (M.Sc. Thesis) Department of Biology, Chulalongkorn University, Bangkok. 105 pp.

Leh, C.M.U., S.K. Poon and Y.C. Siew (1985). Temperature-related Phenomena affecting the Sex of Green turtle *(Chelonia mydas)*

Hatchlings in the Sarawak Turtle Islands. *Sarawak Museum Journal* 34(55): 183–193.

Liew, H.C., E.H. Chan, F. Papi and P. Lutschi (1996). Long Distance Migration of Green Turtles from Redang Island, Malaysia: The Need for Regional Cooperation in Sea Turtle Conservation. B. Devaux (ed.). Pp. 73-75. In: *Proceedings of the International Congress of Chelonian Conservation.* Editions Soptom, Gonfaron.

Lim, B.L., L. Ratnam and N.A. Hussein (1995). Turtles from the Sungei Singgor Area of Temengor Forest Reserve, Hulu Perak, Malaysia. *Malayan Nature Journal* 48: 365–369.

Lim, K.K.P. and F.L.K. Lim (1992). *A Guide to the Amphibians and Reptiles of Singapore.* Singapore Science Centre, Singapore. 160 pp.

Loch, J.H. (1951). Notes on the Perak River Turtle. *Malayan Nature Journal* 5: 157–160.

Louman, J.W.W. (1982). Breeding the Six-footed Tortoise *Geochelone emys* at Wassenaar Zoo, Netherlands. In: P.J.S. Olney (ed). Pp. 153-156. *International Zoo Yearbook 1982.* Zoological Society of London, London.

Luxmoore, R. (1996). The Trade in Reptiles. In: T. Whitten and J. Whitten (eds). *Indonesian heritage. Wildlife. Volume 5.* Pp. 106–107. Editions Didier Millet/Archipelago Press, Singapore.

Lyons, D.D. (1969). Keys to Shell-bones of Bornean Emydid Turtles. *Sarawak Museum Journal* 17: 89–95.

McKeown, S., D.E. Meier and J.O. Juvik (1990). The Management and Breeding of the Asian Forest Tortoise (*Manouria emys*) in Captivity. Pp. 138–159. In: *Proceedings First International Symposium on Turtles and Tortoises: Conservation and Captive*

Husbandry, 1990. California Turtle and Tortoise Society, Orange, CA.

McDowell, S.B. Jr. (1964). Partition of the Genus *Clemmys* and Related Problems in the Taxonomy of the Aquatic Testudinidae. *Proceedings of the Zoological Society of London* 143(2): 239–279.

Manthey, U. and W. Grossmann (1997). *Amphibien and Reptilien Sudostasiens.* Natur und Tier, Münster. 512 pp.

Mehrtens, J.M. (1970). *Orlitia* the Bornean Terrapin. *International Turtle & Tortoise Society Journal* 4(5): 6–7, 33.

Mertens, R. (1937). Ueber *Callagur borneoensis* (Schlegel & Müller), eine seltene Brackwasser-Schildkröte. *Aquarien und Terrarien* (7): 150–151.

Mertens, R. (1942). Zwei Bemerkungen über Schildkröten Sudost-Asiens. I. Über einen bemerkenswerten Geschlechtsunterschied bei *Geoemyda spinosa* (Gray). *Senckenbergiana* 25(1/3): 41–46.

Mertens, R. (1971). Die Stachelschildkröte (*Heosemys spinosa*) und ihre Verwandten. *Salamandra* 7(2): 49–54.

Meylan, A. (1988). Spongivory in Hawksbill Turtles: A Diet of Glass. *Science* 239: 393–395.

Meylan, P.A. (1987). The Phylogenetic Relationships of Soft-shelled Turtles (Family Trionychidae). *Bulletin of the American Museum of Natural History* 186(1): 1–101.

Moll, E.O. (1978). Drumming along the Perak. *Natural History* 87(5): 36–43.

Moll, E.O. (1980). Tuntong Laut, the River Turtle that goes to Sea. *Nature Malaysiana* 5(2): 16–21.

Moll, E.O. (1984). River Terrapin Recovery Plan for Malaysia. *The Journal of Wildlife and Parks* 3: 37–47.

Moll, E.O. (1985). Estuarine Turtles of Tropical Asia: Status and Management. In: *Proceedings Symposium Endangered Marine Animals and Marine Parks*, 1985(1): 214–226.

Moll, E.O. (1989). New Protection for Malaysian Turtles. *Tortoises and Turtles (IUCN T&FWTSG Newsletter)* (4): 8–9.

Moll, E.O. (1989). Tortoises of Tropical Asia: Regional Introduction. Pp. 111–112; *Indotestudo elongata*, Elongated Tortoise. Pp. 116–117. *Manouria emys*, Asian Brown Tortoise. Pp 119–120. *Manouria impressa*, Impressed Tortoise. P. 121. In: *The Conservation Biology of Tortoises*. I.R. Swingland and M. W. Klemens (eds.). Occasional Papers IUCN Species Survival Commission (5). IUCN, Gland.

Moll, E.O. (1980). Natural History of the River Terrapin, *Batagur baska* (Gray) in Malaysia (Testudines: Emyidae). *Malaysian Journal of Science* 6A: 23–62.

Moll, E.O. (1985). Comment: Sexually Dimorphic Plastral Kinesis—the Forgotten Papers. *Herpetological Review* 16(1): 16.

Moll, E.O. (1990). *Status and Management of the River Terrapin* (Batagur baska) *in Tropical Asia*. Final report to the World Wildlife Fund-U.S. (Project 3901), Washington, D.C.

Moll, E.O. and M. Khan bin M. Khan (1990). Turtles of Taman Negara. *Journal of Wildlife and Parks* 10: 135–138.

Morreale, S.J., G.J. Ruiz, J.R. Spotila and E.A. Standora (1982). Temperature-dependent Sex Determination: Current Practise Threaten Conservation of Sea Turtles. *Science* 216: 1245–1247.

Mrosovsky, N. (1981). Plastic Jellyfish. *Marine Turtle Newsletter* (17): 5–7.

Mudde, P. (1981). De Amboinese Doorsschildpad (*Cuora amboinensis*). *Lacerta* 39(6 & 7): 65–67.

Mudde, P. (1982). *Cuora amboinensis* de Amboinese Doorsschildpad. *Lacerta* 40(10 & 11): 261–263.

Nair, P.N.R. and M. Badrudeen (1975). On the Occurrence of the Softshell Turtle *Pelochelys bibroni* (Owen) in Marine Environment. *Indian Journal of Fisheries* 22: 270–274.

Ng, P.K.L. and Y.C. Wee (1994). *The Singapore Red Data Book. Threatened Plants and Animals of Singapore.* The Nature Society (Singapore), Singapore. 343 pp.

Nutaphand, W. (1979). *The Turtles of Thailand.* Siamfarm Zoological Garden, Bangkok. 222 pp.

Nutaphand, W. (1986). ["Manlai"; the World's Biggest Softshell Turtle.] *Thai Zoological Magazine, Bangkok* 1 (4): 64–70. [In Thai.]

Nutaphand, W. (1990). [Softshell Turtles]. *Thai Zoological Magazine, Bangkok* 5 (56): 93–104. [In Thai.]

Obst, F.J. (1983). Beiträg zur Kenntnis der landschildkröten-Gattung *Manouria* Gray, 1852 (Reptilia, Testudines, Testudinidae). *Zoologisches Abhandlungen des Staatliches Museums fur Tierkunde, Dresden* 38(15): 247–256.

Ong, J.E. (1975). Turtles in Penang. *Malaysian Naturalist* 2(1): 5–6.

Pan Khang Aun (1990). Malayan Turtles. *Journal of Wildlife and Parks.* 9: 20–31.

Polunin, N.V.C. (1975). *Sea Turtle Reports on Thailand, West Malaysia, and Indonesia with a Synopsis of Data on the Conservation Status in the Indo-West Pacific Region.* Report to the IUCN, Morges. 40 pp.

Pough, F.H., R.M. Andrews, J.E. Cadle, M.L. Crump, A.H. Savitzky and K.D. Wells (1998). *Herpetology*. Prentice Hall, Upper Saddle River, New Jersey. xi + 577 pp.

Pritchard, P.C.H. (1979). *Encyclopedia of Turtles*. T.F.H. Publications, Inc. Ltd., Neptune, New Jersey. 895 pp.

Pritchard, P.C.H. (1984). Piscivory in Turtles and Evolution of the long-necked Chelidae. *Symposium of the Zoological Society of London* 52: 87–110.

Pritchard, P.C.H. (1993). Carapacial Pankinesis in the Malayan Softshell Turtle, *Dogania subplana*. *Chelonian Conservation and Biology* 1(1): 31–36.

Rhodin, A.G.J., R.A. Mittermeier and P.M. Hall (1993). Distribution, Osteology, and Natural History of the Asian Giant Softshell Turtle, *Pelochelys bibroni*, in Papua New Guinea. *Chelonian Conservation and Biology* 1: 19–30.

Rogner, M. (1995). *Schildkröten 1. Chelydridae Dermatemydidae Emydidae*. Heidi Rogner-Verlag, Hürtgenwald. 192 pp.

Rogner, M. (1996). *Schildkröten 2. Kinosternidae Platysternidae Testudinidae Trionychidae Carettochelyidae Cheloniidae Dermochelyidae Chelidae Pelomedusidae*. Heidi Rogner-Verlag, Hürtgenwald. 265 pp.

Rummler, H.-J. and Fritz, U. (1991). Geographische variabilität der Amboina-Scharnierschildkröte *Cuora amboinensis* (Daudin, 1802), mit Beschreibung einer neuen unterart, *C. a. kamaroma* subsp. nov. *Salamandra* 27(1): 17–45.

Sachsse, W. (1971). Beobachtungen an Jungen *Chitra indica* Insbesondere zum Beuteerwerb. *Salamandra* 7: 31–37.

Sharma, D.S.K. (1996). Conservation and Management Project for Endangered Estuarine Turtles in Peninsular Malaysia. In: B.

Devaux (ed.). Pp. 76. *Proceedings of the International Congress of Chelonian Conservation.* Gonfaron, France, 6–10 July 1995. Editions SOPTOM, Gonfaron.

Siow, K.T. and E.O. Moll (1982). Status and Conservation of Estuarine and Sea Turtles in West Malaysian Waters. In: K. A. Bjorndal (ed.). *Biology and Conservation of Sea Turtles.* Pp. 339–347. Smithsonian Institution Press, Washington, D.C.

Smith, M.A. (1922). On a Collection of Reptiles and Batrachians from the Mountains of Pahang, Malay Peninsula. *Journal of the Federated Malay States Museums* 10 (4): 263–282.

Smith, M.A. (1930). The Reptilia and Amphibia of the Malay Peninsula from the Isthmus of Kra to Singapore, including the Adjacent Islands. *Bulletin of the Raffles Museum, Singapore, Straits Settlements* (3): 1–138.

Smith, M.A. (1931). *Fauna of British India, including Ceylon and Burma. Vol. I. Loricata, Testudines.* Taylor and Francis, London. xxviii + 185 pp + 2 pl.

Swindells, R.J. and F.C. Brown (1964). Ability of *Testudo elongata* Blyth to withstand excessive heat. *British Journal of Herpetology* 3: 166.

Taylor, E.H. (1970). The Turtles and Crocodiles of Thailand and Adjacent Waters. *University of Kansas Science Bulletin* 49(3): 87–179.

Thirakhupt, K. and P.P. van Dijk (1995 "1994"). Species Diversity and Conservation of Turtles of Western Thailand. *Natural History Bulletin of the Siam Society* 42: 207–259.

Tweedie, M.W.F. (1953). The Breeding of the Leathery Turtle. *Proceedings of the Zoological Society of London* 123(2): 273–275.

Uchida, I. (1979). *[Brief Report on the Hawksbill Turtle* (Eretmochelys imbricata) *in the Waters adjacent to Indonesia and Malaysia.]* Japanese Tortoise Shell Society, Nagasaki. [In Japanese.]

Uchida, I. (1980). *[The Report of a Feasibility Research on Artificial Hatchery and Cultivation of Hawksbill Turtle* Eretmochelys imbricata *in Waters adjacent to Malaysia, Singapore and Indonesia].* Japanese Tortoise Shell Society, Nagasaki. [In Japanese.]

Vijaya, J. (1982). *Pelochelys bibroni* from the Gahirmatha Coast, Orissa. *Hamadryad* 7(3): 17.

Webb, R.G. (1995). Redescription and Neotype Designation of *Pelochelys bibroni* from Southern New Guinea (Testudines: Trionychidae). *Chelonian Conservation and Biology* 1: 301–310.

Webb, R.G. (1997). Geographic Variation in the Giant Softshell turtle, *Pelochelys bibroni*. Linneaus Fund Research Report. *Chelonian Conservation and Biology* 2(3): 450.

Wermuth, H. and R. Mertens (1977). *Liste der rezenten Amphibien und Reptilien. Testudines, Crocodylia, Rhynchocephalia.* Das Tierreich 100. Walter de Gruyter & Co. 174 pp.

World Wide Fund for Nature. (1997). *Review of Legislation affecting Marine and Freshwater Turtle, Terrapin and Tortoise Conservation and Management in Malaysia: Recommendation for Change.* Report produced under Project MYS 343/96, January, 1996. Kuala Lumpur.

Wyatt-Smith, J. (1960). The Conservation of the Leathery Turtle, *Dermochelys coriacea. Malayan Nature Journal* 14: 194–199.

Witzell, W.N. (1983). Synopsis of Biological Data on the Hawksbill Turtle, *Eretmochelys imbricata* (Linnaeus, 1766). *FAO Fisheries Synopsis* (137): i–iv + 1–78.

Youngprapakorn, P. (1991). First Observation on Six-legged Turtle's egg-laying. *Feature Magazine, Bangkok* (80): 148–158. [in Thai.]

Zug, G.R. (1993). *Herpetology, An Introductory Biology of Amphibians and Reptiles.* Academic Press, San Diego. xv + 527 pp.

Index

Compiled by C.L. Chan

(Page numbers in bold indicate illustrations)

A

abdominal 2, **10**
aboriginal people 7
Acrochordus javanicus 84
Africa south of the Sahara 2
Agal 15
aggressive species 39
agricultural fields 6
agricultural pesticides 81
agriculture sites 92
Aldabra 3
Aldabra Tortoise **xii**
algae 28, 32, 34, 46
alligators 1
Amboina 63
Ambon 63, 65
ambush feeder 48
American Hardshell Turtles 93
American markets 113
Amur River 51
Amyda cartilaginea **7**, 8, 12, 38, 113
Amyda cartilaginea, close-up of head **39**
anal 2, **10**
animal flesh 51
animal kills 102
animal matter 72, 84
animated pancakes 37
anterior 118
aphrodisiac properties 19, 113
aquaria 113
aquatic 72, 93, 118

aquatic animals 46
aquatic insects 39, 65
aquatic invertebrates 28
aquatic mode of life 115
aquatic organisms 6
aquatic plants 48, 65
arribada 118
'arribadas' 34
Asian box turtles 66
Asian Brown Tortoise 4, 5, 9, 99, 103, **104**, 107, **107**, 109
Asian Brown Tortoise, close-up of head **103**
Asian Brown Tortoise, carapace **106**
Asian Brown Tortoise, close-up of scales on leg **105**
Asian Brown Tortoise, close-up of spur on hind leg **105**
Asian Brown Tortoise, lateral view of shell **106**
Asian Brown Tortoise, plastron **106**
Asian Brown Tortoise, ventral view **108**
Asian Giant Softshell Turtle 9, 38, 46, **47**, 48
Asian Giant Turtle 9
Asian Hardshell Turtles 53
Asian Leaf Turtle 9, 53, 66, **67**, **68**
Asian Leaf Turtle, plastron **67**, **69**

Index

Asian mainland   48
Asian Softshell Turtle   **7**, 8, 38, 113
Asian Softshell Turtle, close-up of head   **39**
Australasia   2
Australia   4, 22, 96
Axillary   **10**
Ayeyarwady River   41

B

bamboo forest   112
bamboo shoots   109
bananas   89, 90
Banding lontong   99
Bangkok   78
Bangladesh   41, 42, 48, 55, 65, 109
Bangladesh, eastern   101
Bangladesh, north-eastern   107
Bangladesh, south-eastern   107
Baning dayak   85, **114**
Baning perang   103
Banning bukit   110
Batagur baska   **vii**, 8, 9, 13, 53, 54, **57**, 113
Batagur baska, hatchling   **54**
Batagur baska, male in non-breeding colour   **56**
Batagur baska, Male in breeding colour   **56, 57**
Bataguridae   9, 53
batagurids   93
Batang Ai National Park   **36, 86–87**
bays   32
beak   118
beef   40
beef heart   89

behaviour   8
Beluku   58, 66
biology   8
bioluminescence   19
Biuku   58, 81
Black Marsh Turtle   **3**
Black Pond Turtle   9, 54, 90, **91**, 92
Black Pond Turtle, close-up of head   92
Black Pond Turtle, plastron   **91**
Borneo   24, 29, 38, 39, 46, 48, 58, 62, 65, 66, 68, 73, 82, 85, 89, 92, 107, 114
Bota Kanan   58
bottom-living organisms   23
bottom-walkers   53, 68, 81, 92
boxes   33
brackish waters   8
breeding behaviour   42
breeding season   58
bridge   2
British East India Company   46
broad-leaved forests   107
Brotia costula   81
Brunei Darussalam   7, 29
Buddhism   78
Buddhist inscriptions   109
Bukit Pinang   58
Bukit Timah Nature Reserve   77
Bulus   38, 66
Burma   113

C

Callagur borneoensis   9, 13, 53, 58, 113
Callagur borneoensis, adult female   **61**

Callagur borneoensis, adult male 59
Callagur borneoensis, adult male in partial breeding colour 59
Callagur borneoensis, close-up of the head 60
Callagur borneoensis, hatchling 61
Cambodia 38, 55, 72, 77, 80, 92
Cameron Highlands 112
canals 52, 65, 80
canned dog food 89
carapace 2, **10**, 118
Caretta caretta 8, 11, 21–23,
carnivorous 28, 38, 42
carrion 39, 51, 68, 98
cartilage 6
catchment areas 65
Central America 53, 96
South America 53
Central Catchment Area 77
Ceram 65
Checklist 8
Chelodina rugosa 4
Chelonia mydas 8, 11, 21, 24, **24, 27**
Chelonia mydas, hatchlings **28**
Chelonia mydas, laying eggs **24**
Chelonia mydas, nesting **24**
Chelonia mydas, egg **29**
Chelonia olivacea 34
Chelonii 1
Cheloniidae 8, 15, 21
chicken 40, 109
China 6, 49, 51, 112
China, southern 22, 48, 107
Chinese Softshell Turtle 9, 38, 46, 49, **49**

Chinese Softshell Turtle, close-up of head **50**
Chinese Softshell Turtle, hatchling **51**
Chinese Softshell Turtle, hatchling plastron **52**
Chitra 40, 41, 47
Chitra chitra 40, 41, **41,** 42
Chitra indica 40–43
Chitra indica species-complex 8, 12, 38, 40
circumtropical 118
CITES 42, 46, 48, 52, 58, 62, 102, 109, 112, 118
city parks 81
city ponds 96
clams 62
clear forest stream **4**
cloaca 92
clutch size 118
coastal areas 55
coastal people 58
coastal regions 48
coastal waters 8
Coleroon 41
collection for food 115
collection for medicine 115
combs 33
connoisseur fare 113
conservation 113, 115
conservation status 8
coral cervices 32
coral reefs 32
Cosmopolitan 16, 32
Costals 2, **10**
courtship 118
crabs 23, 39, 48, 51, 81
crepuscular 68, 75, 118

Index

crickets 84
'Critically Endangered' 42, 62
crocodiles 1, 5, 7
crocodilians 1
crustaceans 19, 32, 34, 55, 65
Cryptodira 1
Cryptodirans 1–3
ctenophores 19
Cuora 66
Cuora amboinensis 9, 13, 53, 63, **64**, 113
Cuora amboinensis, close-up of head **63**
Cuora amboinensis, plastron **64**
Cuora amboinensis kamaroma 63, 65
curios 33
Cyclemys annandalei 77
Cyclemys dentata 9, 13, 53, 66, **67**, 68, **68**, 69
Cyclemys dentata complex 68
Cyclemys dentata, plastron **67**, **69**
Cyclemys oldhamii 68

D

'Data Deficient' 84
daybreak 35
deciduous forests 101
deep ponds 92
deforestation 77, 114, 115
Department of Wildlife and National Parks 58, 115
depressed 118
dermal 118
Dermochelyidae 8, 15
Dermochelys coriacea 2, 8, 11, 15, **17**

Dermochelys coriacea, hatchling **18**
developmental projects 40
diet 8, 68
dietary requirements 42
Dillenia species 101
dimorphic 118
Dipsochelys gigantea **xii**
disc 118
diseases 6
distribution 8
Dogania subplana 9, 12, **37**, 38, 43, **44**, 113
Dogania subplana, close-up of head **43**
Dogania subplana, juvenile **45**
dorsum 118
drugs 6
earthworms 84
East Asia 52, 109
east coast 29
East Malaysian States 29
ecosystems 115
Eichhornia crassipes 6
Elephant Trunk Snake 84
Elongated Tortoise 9, 99, **100**, 101
Elongated Tortoise, close-up of head **102**
Elongated Tortoise, plastron **101**
Emydidae 9, 93
Emys baska 54
Emys borneoensis 58
Emys crassicollis 90
Emys dentata 66
Emys platynota 81
Emys spinosa 72
Emys subtrijuga 79

environment 115
Eretmochelys imbricata 8, 11, 21, 30, **31–33**
Eretmochelys imbricata, Close-up of head **30**
estuaries 32, 55, 62, 89
estuarine areas 78
ethnic Chinese 40
Europe 65
European markets 113
evergreen forest 112
extinction 6

F
fallen figs 109
fangs 5
fat 6
Fatehgarh 43
feeding behaviour 42
feeding trough 109
femoral 2, **10**
Ficus racemosa 101
figs 68, 72
filaria 6
Filopaludina sumatrensis 81
fish 34, 75, 89
Fisheries Department 115
fishes 23, 39, 42, 46, 48, 51, 55, 81
fishing gear 66, 114
flippers 4
food 40, 109
forelimbs 4
forest fires 109, 114
forest stream **36**
forest-adapted animals 115
forest-adapted plants 115
forest-dwelling species 114

forested areas 101
forested land 114
forests of Borneo 113
forests of Peninsular Malaysia 113
France 96
Frazer's Hills 112
freshwater bodies 39
freshwater marshes 65
Freshwater mussels 81
Freshwater snails 81
Freshwater turtles 4, 42
freshwaters 8, 78, 119
frogs 51, 92, 109
fruits 39, 46, 48, 62, 78, 84, 98, 101
fungi 65, 101

G
Galapagos 3
gall bladder 6
Ganga 41
Ganga River 43
genotypic sex determination mechanism 4
Geoemyda grandis 70
Geoemyda impressa 110
Geoemyda spengleri 9
giant turtles 41
Glossary 118
Gmelina arborea 101
Godavari 41
Green Turtle 8, 21, 24, **24**, **27**, 28, 29, 32
Green Turtle, hatchlings **28**
Green Turtle, laying eggs **24**
Green Turtle, nesting **24**
Green Turtle, egg **29**

Index

ground-dwelling turtles   114
Gular   **10**
gular   2

H

habitat alternation   40
habitat change   115
habitat destruction   109, 112
Hainan   112
hard-bodied molluscs   21
hardwoods   114
harpoon guns   114
Hawaii   51, 96
Hawksbill   32
Hawksbill Sea Turtle   8, 21, 30, **31–33**
Hawksbill Sea Turtle, Close-up of head   **30**
Heosemys grandis   9, 14, 53, 70
Heosemys grandis, adult   **71**
Heosemys grandis, close-up of head   **70**
Heosemys grandis, juvenile   **71**
Heosemys spinosa   9, 14, 53, 72, **73**
Heosemys spinosa, adult   **76**
Heosemys spinosa, carapace   **74**
Heosemys spinosa, juvenile plastron   **75**
Heosemys spinosa, plastron   **74, 76**
herbicide   6
herbivores   99
herbivorous   72, 75, 78, 84, 109
Hieremys annandalei   9, 14, 53, 77, **78**
hill forests   68, 73
Homalopsis buccata   89

horsemeat   89
human habitated areas   92
human habitated environments   65
human induced threat   6
Humeral   2, **10**

I

imbricate   119
Impressed Tortoise   9, 99, 110, **110**, 111, 112
Impressed Tortoise, close-up of head   **112**
Impressed Tortoise, plastron   **111**
Incubation period   4
India   34, 40–42, 48, 96, 113
India, eastern   48, 55
India, north-eastern   65, 68, 101, 107
India, northern   42, 43, 101
Indian freshwater turtle   79
Indian Narrow-headed Softshell Turtle   40
Indian Sunderbans   55
indigenous forest people   77
indigenous people   69, 113
Indo-China   65, 107
Indo-Pacific   34
Indonesia   28, 48, 51, 55, 63, 73
Indotestudo elongata   9, 12, 99, **100**
Indotestudo elongata, close-up of head   **102**
Indotestudo elongata, plastron   **101**
Indus   41
Industrialisation   114
inframarginal pore   119

Inguinal **10**, 119
insect larvae 81
insects 5, 51, 75
'Insufficiently Known' 109
invertebrate prey 34
invertebrates 32, 68, 92, 109
Ipomoea aquatica 89
irrigation canals 39, 78
irrigation reservoirs 96

J

Japan 22, 51, 96, 107
Java 39, 41, 46, 65, 68, 80, 82, 92
Jelebu hitam 90
Jelebu siput 79
jellyfish 3, 18, 19, 34
jellyfish-eaters 15
jewellery 33
Juku juku besar 85
juvenile 119

K

Kadazandusun 103
Kalimantan 48
Kamban 15
Kanburi Narrow-headed Softshell Turtle 40, 41, **41**
Kangkong 89
Karimundjawa 41
Kedah 58, 80
keel 119
Kembau 15
Key to Species 11
Klang River 89, 90
Korea 51
Kuala Behrang 58
Kuala Tahan River 41, 42

Kuja 38
Kura 66
Kura besar 70
Kura Katap 63
Kura kura 63, 66, 72
Kura kura anam kaki 103
Kura kura duri bukit 72
Kura kura kolam 90
Kura kura mas 99
Kura kura patah 63
Kura punggung datar 81
Kura terlinga-merah 93
Kura tokong 77

L

Labi besar 46
Labi biasa 38
Labi bunga 40
Labi china 49
Labi labi 38
Labi melayu 43
lagoons 32
lakes 89, 96
Land Tortoises 99
Langkawi Islands 46
Laos 38, 72, 80, 112
large river systems 62
large rivers 48, 89
largest freshwater turtle 85
Late Triassic 2
law makers 115
Layang Layang **24**
Layi layi 43
leaf litter 75, 109
Leatherback Sea Turtle 2, 3, 8, 15, **17**, 19, 32
Leatherback Sea Turtle hatchling **18**

leaves 101
Lepidochelys 3
Lepidochelys olivacea 8, 11, 21, 34, **35**
lizards 1, 5
local cultures 7
locked-up nutrients 115
Loggerhead Sea Turtle 8, 21, 22, 23, **23**
logging 114
Logging trails 114
Lombok 39
London 43
'Lower Risk: Near Threatened' 66, 72, 90, 98
lower shell 2
lowland forests 65, 73
lowland rainforest 84

M

macrophytes 78
Mae Klong river basin 41
Mahanadi 41
malaria 6
Malay *Beluku* 58
Malay Peninsula 42, 68, 109
Malayan Box Turtle 9, 53, 63, **64**, 113
Malayan Box Turtle, close-up of head **63**
Malayan Box Turtle, plastron **64**
Malayan Flat-shelled Turtle 9, 54, 81
Malayan Flat-shelled Turtle, adult **83**
Malayan Flat-shelled Turtle, close-up of head **83**
Malayan Flat-shelled Turtle, hatchling **82**
Malayan Flat-shelled Turtle, juvenile **84**
Malayan Giant Turtle 54, 85, **85**, **86–87**, **89**
Malayan Giant Turtle, close-up of head **88**
Malayan Snail-eating Turtle 9, 53, 79, **79**
Malayan Snail-eating Turtle, close-up of head **80**
Malayan Softshell Turtle 9, **37**, 38, 43, **44**, 45, 113
Malayan Softshell Turtle, close-up of head **43**
Malayan Softshell Turtle, juvenile **45**
Malayemys subtrijuga 9, 13, 53, 79, **79**
Malayemys subtrijuga, close-up of head **80**
Malaysia 48, 51, 96
Maluku 63
mandible 119
mangrove areas 34, 55
mangrove forests 62
mangrove fruits 62
mangrove leaves 62
mangrove plant 55
mangrove swamps 65
Manouria emys **viii**, **x**, 4, 9, 12, 99, 103, **104**, **107**
Manouria emys, carapace **106**
Manouria emys, close-up of head **103**
Manouria emys, close-up of scales on leg **105**
Manouria emys, close-up of spur on hind leg **105**

Manouria emys, lateral view of shell **106**
Manouria emys, plastron **106**
Manouria emys, ventral view **108**
Manouria emys emys 107
Manouria emys phayrei 107
Manouria impressa 9, 12, 99, 110, **110**
Manouria impressa, close-up of head **112**
Manouria impressa, plastron **111**
Map of Borneo **iii**
Marginals 2, **10**, 119
marine 119
marine animal 7
marine invertebrates 18
marine pollution 19
markets, Peninsular Malaysia 113
markets, Sabah 113
markets, Sarawak 113
marshes 51, 72, 80, 84, 92, 96
Maxwell's Hills 112
mealworms 84
meat 6, 75
medicinal properties 19, 113
medicinal qualities 40, 109
medicine 40, 109
medusae 19
Melanochelys trijuga 79
Mergui Archipelago 46
migration 119
Mindanao 73
Missisippi River 96
molluscs 23, 32, 34, 42, 46, 48, 55, 65
monitor lizards 7
mosquito larvae 6

mosquitoes 6
mudbanks 51
muddy rivers 39
mushrooms 102, 109, 112
Myanmar 28, 38, 41, 48, 55, 65, 68, 101, 107
Myanmar, northern 112
Myanmar, southern 72, 73, 92
Myanmar, western 112

N

Narrow-headed Softshell Turtle 8, 38, 40, 41, 47
Natunas 73
Nepal 41, 68
Nepal, southern 101
Nesting 34
Nesting beaches 34
nesting habits 4
nesting season 19
New Guinea 48
New Zealand 1
Nicobars 65
nocturnal 51, 119
nocturnal activity 34
North America 96
North American family 93
Northern Long-necked Turtle 4
Notochelys platynota 9, 13, 54, 81
Notochelys platynota, adult **83**
Notochelys platynota, close-up of head **83**
Notochelys platynota, hatchling **82**
Notochelys platynota, juvenile **84**
Nuchal **10**, 119
nuchal scale 2

Nymphaea spp. 78

O

odious secretion 92
offensive odour 92
Oil 19
oil palm 65
Old World 53
Olive Ridley Sea Turtle 8, 21, 22, 34, 35, **35**, 48
omnivore 92
omnivorous 32, 84, 89, 96
Orang Asli 77
Orang Asli villages 109
Orange-headed Temple Turtle 9, 53, 70
Orange-headed Temple Turtle, adult **71**
Orange-headed Temple Turtle, close-up of head **70**
Orange-headed Temple Turtle, juvenile **71**
orchards 65
Orissa 101
Orlitia borneensis 9, 13, 54, 85, **85**, **86–87**, **89**, **114**
Orlitia borneensis, close-up of head **88**
otters 7
overharvest of eggs 115
ox-bow lakes 48

P

Padma 41
Pahang 41, 112
Painted Terrapin 9, 53, 58, 62, 113
Painted Terrapin, adult female **61**
Painted Terrapin, adult male **59**
Painted Terrapin, adult male in partial breeding colour **59**
Painted Terrapin, close-up of head **60**
Painted Terrapin, hatchling **61**
Pakistan 41
papayas 89
parietal 119
peat swamps 92
pectoral 2, **10**
Pelochelys 48
Pelochelys bibroni 48
Pelochelys cantorii 9, 12, 38, 46, **47**, 48
Pelodiscus sinensis 9, 38, 46, 49, **49**
Pelodiscus sinensis, close-up of head **50**
Pelodiscus sinensis, hatchling **51**
Pelodiscus sinensis, plastron of hatchling **52**
Penang 43, 78
Peninsular Malaysia 19, 21, 22, 29, 34, 38, 40, 48, 52, 55, 58, 62, 63, 65, 66, 72, 73, 80, 81, 89, 92, 96, 103, 107, 109, 112, 113, 114
Peninsular Malaysia, northern 77
Penyu agar 24
Penyu belimbing 15
Penyu emegit 24
Penyu hijau 24
Penyu karah 22, 30
Penyu kepala besar 22
Penyu lilin 30
Penyu lipas 34
Penyu pulau 24

Penyu sisek 30
Penyu sisek tempurong 22
Penyu timbo 15
Perak 58, 78
Perak River 55
Perlis 80
pest 96
Pesticide 6
pet trade 66, 69, 77, 93, 96, 113, 115
pet turtle 65, 98
Petani River 80
pets 62, 77
Philippines 28, 65, 73
pig 37, 109
plains 68
plankton 18
planners 115
plant matter 38
plantations 65
plants 65, 68, 92
plastic bags 19
plastral concavity 107
plastron 2, **10**, 19
Pleurodira 1
Pleurodirans 2
polluted areas 114
pollution 42
pollution of waterways 66
polygonal scales 16
ponds 6, 39, 51, 65, 68, 78, 84
pools 81
poor swimmers 75
pork 40
powerful jaws 39
prawns 39, 46
pre-World War II 58
prefrontals 119
private collectors 77

proboscis 119
Professor John Leonard Harrison v
proteinaceous food 6
Puff-faced Water Snake 89
Pulau Pinang 43, 48
Pulau Selingan **20**, **24**
pyrosomas 19, 119

R

rachemys scripta, adult **94**
Rantau Abang beach **17**, 19
rays 48
Red-eared Slider 9, 93, **94**
Red-eared Slider, plastron of adult **95**
Red-eared Slider, basking adult **96**
Red-eared Slider, close-up of head **97**
Red-eared Slider, hatchling **97**
Red-eared Slider, plastron of juvenile **98**
reproduction 8
reptile groups 115
reservoirs 48, 51
restauranteurs 40
restaurants 113
revenue earning activities 115
rice fields 65, 80
"ricefield turtle" 80
Ridley Sea Turtles 3
River Terrapin 8, 9, 53–55, 58, 113
rivers 6, 51, 62, 72, 78
Royal College of Surgeons of England 43
rubber 65
Russia 51

S

Sabah 29
salinity 119
sand mining 42
sandy beaches 34
Sarawak 29
sargassum 28
schistasomiasis 6
Schyphomedusidae 18
scute 119
sea anemones 32
sea grasses 28
sea urchins 23
sea weeds 21
seafarers 29
seagrass beds 28
seedlings 109
seeds 39
Seeds of plants 51
Semi-aquatic 65
serrated 119
sexual dimorphism 8
Seychelles 1
shallow ponds 52
shallow waters 84
sharks 48
sharp beak 39
shell ramming 102
shellfish 62
shoots 101
Shorea sp. 102
shrimps 48, 81, 92
side-necked turtles 2
Siebenrockiella crassicollis **3**, 9, 13, 54, 90, **91**
Siebenrockiella crassicollis, close-up of head 92
Siebenrockiella crassicollis, plastron **91**
Singapore 7, 22, 23, 38, 39, 40, 42, 46, 52, 55, 63, 72, 73, 77, 81, 82, 92–94, 96
slow flowing larger streams 68
slow flowing rivers 80
sluggish streams 92
slugs 101
small animals 98
small rivers 96
small streams 68
small waterfalls 68
snail eaters 6
snails 6, 51, 84, 92
snakes 1, 5
soft fruits 89
softshell turtles 4, 37, 40, 113
Solo River 41
Sonneratia 55
South Africa 2, 96
South America 2
south-east Asia 41, 65, 107, 113
southern Asia 41, 42
specialized habitat 42
Spiny Hill Turtle 9, 53, 72, **73**
Spiny Hill Turtle, adult **76**
Spiny Hill Turtle, carapace **74**
Spiny Hill Turtle, juvenile plastron **75**
Spiny Hill Turtle, plastron **74**, **76**
sponges 21, 23, 32
Sri Lanka 48
streams 65, 72, 84, 107
subtropical seas 32
Sulawesi 65
Sulu Archipelago 73
Sumatra 39, 41, 46, 55, 62, 65, 68, 73, 80, 82, 89, 92, 107
sunset 35
supracaudal 119, 120

Sutong 58
Suyan 103
swamps 78, 84
swimmer 81
symbiotic fish 19

T

taboo 7
Taiwan 51
temple ponds 72, 78, 81, 96
temples 7, 72, 78, 109
terrapin 7
Testudines 1
Testudinidae 9, 99
Testudo amboinensis 63
Testudo caretta 22
Testudo cartilaginea 38
Testudo coriacea 15
Testudo elongata 99
Testudo emys 103
Testudo imbricata 30
Testudo mydas 24
Testudo scripta 93
Testudo scripta elegans 93
Thailand 28, 38, 41, 42, 48, 52, 55, 65, 68, 72, 73, 77, 78, 80, 81, 96, 101, 107, 112, 113
Thailand, central 41
Thailand, southern 62, 80, 82, 107
Thailand, western 41, 46
Theodore Cantor 46
therapeutic value 6
Thomas Nelson Annandale 77
tortoiseshell 33, 120
totem 7
totemic animal 7
Trachemys scripta 9, 93

Trachemys scripta elegans 14
Trachemys scripta elegans, basking adult **96**
Trachemys scripta, adult plastron **95**
Trachemys scripta, close-up of head **97**
Trachemys scripta, hatchling **97**
Trachemys scripta, juvenile plastron **98**
trawl nets 22
trawling nets 32
Trengganu **17**, 19, 58
trionychid 48
Trionychidae 4, 8, 37
Trionyx (Aspidonectes) sinensis 49
Trionyx indicus 40
Trionyx subplanus 43
tropical seas 32
True land tortoises 99
true swimmers 68, 92
tuatara 1
tubercle 120
tubers 109
tunicates 32
Tuntong laut 54
Tuntong sungei 58
Turtle Islands Park **20, 24**
turtle populations 114
turtle soup 29

U

underwater caves 28
United States 96
unshelled eggs 6
unwanted pets 93
urbanisation 81, 114

USA   65
Uttar Pradesh   101

V

vegetables   48, 62, 78, 84, 89, 90, 98
vegetation-chocked ponds   89
venom   5
ventrum   120
Vertebrals   2, **10**, 120
vertebrates   92
vertical-necked turtles   2
Vietnam   38, 48, 51, 55, 65, 82, 92
Vietnam, northern   51, 112
Vietnam, southern   72 77, 80, 81
villages   92
'Vulnerable'   78, 109, 112

W

Wales   16
water catchment areas   92
water hyacinth   6
water lily   78
water plants   78, 84
water pollution   40
water weeds   6, 65
waterbirds   7
waterbodies   39
waterways   93
West Malaysia   23, 24, 32, 38, 42, 43, 46, 48, 55, 68, 72, 77, 78, 80, 82, 85, 93, 94, 101, 112
West Malaysian population   42
wild meat   113
wilderness areas of Borneo   115
wilderness areas of Peninsular Malaysia   115

wire mesh cages   39
wooded areas   73
wooden boats   19
World War II   43
worms   65, 75. 81, 92
worshippers   78

Y

Yellow-headed Temple Turtle   9, 53, 77, **78**, 109
Yunnan   112

Z

Zoological Survey of India   77
zoos   77, 109, 113, 115

Other titles by *Natural History Publications (Borneo)*

For more information, please contact us at

Natural History Publications (Borneo) Sdn. Bhd.
A913, 9th Floor, Wisma Merdeka
P.O. Box 13908, 88846 Kota Kinabalu, Sabah, Malaysia
Tel: 088-233098 Fax: 088-240768 e-mail: chewlun@tm.net.my

Mount Kinabalu: Borneo's Magic Mountain—an introduction to the natural history of one of the world's great natural monuments *by* K.M. Wong & C.L. Chan

Enchanted Gardens of Kinabalu: A Borneo Diary *by* Susan M. Phillipps

A Colour Guide to Kinabalu Park *by* Susan K. Jacobson

Kinabalu: The Haunted Mountain of Borneo *by* C.M. Enriquez (Reprint)

National Parks of Sarawak *by* Hans P. Hazebroek & Abang Kashim bin Abang Morshidi

A Walk through the Lowland Rainforest of Sabah *by* Elaine J.F. Campbell

In Brunei Forests: An Introduction to the Plant Life of Brunei Darussalam (Revised edition) *by* K.M. Wong

The Larger Fungi of Borneo *by* David N. Pegler

Pitcher-plants of Borneo *by* Anthea Phillipps & Anthony Lamb

Nepenthes of Borneo *by* Charles Clarke

Nepenthes of Sumatra and Peninsular Malaysia *by* Charles Clarke

The Plants of Mount Kinabalu 3: Gymnosperms and Non-orchid Monocotyledons *by* John H. Beaman & Reed S. Beaman

Dendrochilum of Borneo *by* Jeffrey J. Wood

Slipper Orchids of Borneo *by* Phillip Cribb

The Genus Paphiopedilum (Second edition) *by* Phillip Cribb

The Genus Pleione (Second edition) *by* Phillip Cribb

Orchids of Sumatra *by* J.B. Comber

Gingers of Peninsular Malaysia and Singapore *by* K. Larsen, H. Ibrahim, S.H. Khaw & L.G. Saw

Aroids of Borneo *by* Alistair Hay & Peter C. Boyce

Mosses and Liverworts of Mount Kinabalu *by* Jan P. Frahm, Wolfgang Frey, Harald Kürschner & Mario Manzel

Birds of Mount Kinabalu, Borneo *by* Geoffrey W.H. Davison

Aroids of Borneo *by* Alistair Hay & Peter C. Boyce

Mosses and Liverworts of Mount Kinabalu
 by Jan P. Frahm, Wolfgang Frey, Harald Kürschner & Mario Manzel

Birds of Mount Kinabalu, Borneo *by* Geoffrey W.H. Davison

The Birds of Borneo (Fourth edition)
 by Bertram E. Smythies (Revised by Geoffrey W.H. Davison)

The Birds of Burma (Fourth edition) *by* Bertram E. Smythies

Proboscis Monkeys of Borneo *by* Elizabeth L. Bennett & Francis Gombek

The Natural History of Orang-utan *by* Elizabeth L. Bennett

The Systematics and Zoogeography of the Amphibia of Borneo
 by Robert F. Inger (Reprint)

A Field Guide to the Frogs of Borneo *by* Robert F. Inger & Robert B. Stuebing

A Field Guide to the Snakes of Borneo *by* Robert B. Stuebing & Robert F. Inger

The Natural History of Amphibians and Reptiles in Sabah
 by Robert F. Inger & Tan Fui Lian

Marine Food Fishes and Fisheries of Sabah *by* Chin Phui Kong

Layang Layang: A Drop in the Ocean
 by Nicolas Pilcher, Steve Oakley & Ghazally Ismail

Phasmids of Borneo *by* Philip E. Bragg

The Dragon of Kinabalu and other Borneo Stories *by* Owen Rutter (Reprint)

Land Below the Wind *by* Agnes N. Keith (Reprint)

Three Came Home *by* Agnes N. Keith (Reprint)

Forest Life and Adventures in the Malay Archipelago *by* Eric Mjöberg (Reprint)

A Naturalist in Borneo *by* Robert W.C. Shelford (Reprint)

Twenty Years in Borneo *by* Charles Bruce (Reprint)

With the Wild Men of Borneo *by* Elizabeth Mershon (Reprint)

Kadazan Folklore (*Compiled and edited by* Rita Lasimbang)

An Introduction to the Traditional Costumes of Sabah
 (*eds.* Rita Lasimbang & Stella Moo-Tan)

Other titles available through
Natural History Publications (Borneo)

The Bamboos of Sabah *by* Soejatmi Dransfield

The Morphology, Anatomy, Biology and Classification of Peninsular Malaysian Bamboos *by* K.M. Wong

Orchids of Borneo Vol. 1 *by* C.L. Chan, A. Lamb, P.S. Shim & J.J. Wood

Orchids of Borneo Vol. 2 *by* Jaap J. Vermeulen

Orchids of Borneo Vol. 3 *by* Jeffrey J. Wood

Orchids of Java *by* J.B. Comber

A Checklist of the Flowering Plants of Gymnosperms of Brunei Darussalam (eds. M.J.E. Coode, J. Dransfield, L.L. Forman, D.W. Kirkup & Idris M. Said)

Forests and Trees of Brunei Darussalam (eds. K.M. Wong & A.S. Kamariah)

A Field Guide to the Mammals of Borneo *by* Junaidi Payne & Charles M. Francis

Pocket Guide to the Birds of Borneo *Compiled by* Charles M. Francis

The Fresh-water Fishes of North Borneo *by* Robert F. Inger & Chin Phui Kong

The Exploration of Kina Balu *by* John Whitehead (Reprint)

Kinabalu: Summit of Borneo (eds. K.M. Wong & A. Phillipps)

Common Seashore Life of Brunei *by* Marina Wong & Aziah binte Hj. Ahmad

Birds of Pelong Rocks *by* Marina Wong & Hj. Mohammad bin Hj. Ibrahim

Ants of Sabah *by* Arthur Y.C. Chung

Traditional Stone and Wood Monuments of Sabah *by* Peter Phelan

Rafflesia: Magnificent Flower of Sabah *by* Kamarudin Mat Salleh

Borneo: the Stealer of Hearts *by* Oscar Cooke (1991 Reprint)

The Theory and Application of A Systems Approach to Silvicultural Decision Making *by* Michael Kleine

Maliau Basin Scientific Expedition (eds. Maryati Mohamed, Waidi Sinun, Ann Anton, Mohd. Noh Dalimin & Abdul-Hamid Ahmad)

Tabin Scientific Expedition (eds. Maryati Mohamed, Mahedi Andau, Mohd. Nor Dalimin & Titol Peter Malim)

Birds of Pelong Rocks *by* Marina Wong & Hj. Mohammad bin Hj. Ibrahim

Ants of Sabah *by* Arthur Y.C. Chung

Traditional Stone and Wood Monuments of Sabah *by* Peter Phelan

Rafflesia: Magnificent Flower of Sabah *by* Kamarudin Mat Salleh

Borneo: the Stealer of Hearts *by* Oscar Cooke (1991 Reprint)

The Theory and Application of A Systems Approach to Silvicultural Decision Making
   by Michael Kleine

Maliau Basin Scientific Expedition (eds. Maryati Mohamed, Waidi Sinun, Ann Anton, Mohd. Noh Dalimin & Abdul-Hamid Ahmad)

Tabin Scientific Expedition (eds. Maryati Mohamed, Mahedi Andau, Mohd. Nor Dalimin & Titol Peter Malim)

Traditional Cuisines of Sabah

Cultures, Costumes and Traditions of Sabah, Malaysia: An Introduction

Tamparuli Tamu: A Sabah Market *by* Tinar Rimmer

FACE TO FACE WITH NATURE *in* SABAH

Malaysia's premier wildlife conservation attractions are found in Sabah, located in Northern Borneo.
Conservation areas such as the world's largest Orang Utan Rehabilitation Centre at Sepilok, Turtle Islands Park, Gomantong Caves, the rainforest of the Danum Valley, Kinabalu Park and the coral islands of the Tunku Abdul Rahman Park have made Sabah a top nature adventure destination.

SABAH
Malaysian Borneo
The New Millennium Nature
Adventure Destination

For More Information:
SABAH TOURISM PROMOTION CORPORATION
51, Jalan Gaya, 88000 Kota Kinabalu, Sabah, Malaysia.
Tel : (6088) 212121 Fax : (6088) 212075 Email : sabah@po.jarin
Sabah Web site : http://www.sabahtourism.com.my